Manufactured Sites

Manufactured Sites

Rethinking the Post-Industrial Landscape

Edited by
Niall Kirkwood

Taylor & Francis
Taylor & Francis Group

LONDON AND NEW YORK

First published 2001
by Taylor & Francis,
2 Park Square, Milton Park, Abingdon, Oxon, OX14 4RN

Transferred to Digital Printing 2005

Simultaneously published in the USA and Canada
by Taylor & Francis,
270 Madison Ave, New York NY 10016

British Library Cataloguing in Publication Data
A catalogue record for this book is available from the British Library

Library of Congress Cataloging-in-Publication Data

Manufactured sites : rethinking the post-industrial landscape / Niall Kirkwood, editor.
    p. cm.
   Includes bibliographical references and index.
    1. Urban renewal. 2. Brownfields. 3. Urban renewal–Environmental aspects. 4. Urban
landscape architecture. 5. Industrial real estate–Environmental aspects. I. Kirkwood, Niall.

   HT170 .M328 2001
   333.77'15–dc21

                                                  00-046404

   ISBN 0-415-24365-3

Printed and bound by Antony Rowe Ltd, Eastbourne

Contents

Contributors   vii
Foreword by Michael Hough   xii
Preface   xiv

Part I   Introduction   1

**1 Manufactured sites: integrating technology and design
   in reclaimed landscapes**   3
Niall Kirkwood

2 New designs in the legal landscape   12
Rosanna Sattler and David Y. Li with Deborah DiVerdi Carlson and John S. Day

Part II   Integrating technology and design   33
Responses by Rebecca Krinke, Daniel Winterbottom and Niall Kirkwood

**3 Beyond clean-up of manufactured sites: remediation,
   restoration and renewal of habitat**   35
Lucinda Jackson
Response: Remediation, design, and environmental benefit   41

**4 From laboratory to landscape: a case history and possible
   future direction for phyto-enhanced soil bioremediation**   43
Eric Carman
Response: Remediation as engineeering?   50

**5 Phytoremediation: integrating art and engineering through
   planting**   52
Steven Rock
Response: Recovery with plants   59

6 Engineering urban brownfield development: examples from Pittsburgh 61
Sue McNeil and Deborah Lange
Response: Living laboratories: studies in infrastructure and industrial land 69

Contents

7 **The Tacoma Asarco smelter site: the use of geostatistics to guide residential soil clean-up**   72
Dante Tedaldi
Response: Manufacturing site information   *81*

8 **Regeneration: vision, courage and patience**   82
Lorna Walker and Richard Owen
Response: Integration through an interdisciplinary approach?   *102*

9 **Industrial evolution: prevention of remediation through design**   105
Jean Rogers
Response: Re-manufactured sites?   *119*

PART III Reclaimed landscapes   123
Responses by Rebecca Krinke and Daniel Winterbottom

10 **Overview: design practice and manufactured sites**   125
Rebecca Krinke

11 **Landscape Park Duisburg-Nord: the metamorphosis of an industrial site**   150
Peter Latz
Response: Terra-toxic   *162*

12 **Science, engineering, and the art of restoration: two case studies in wetland construction**   166
Wendi Goldsmith
Response: More bio than engineering?   *176*

13 **Fresh Kills landfill: the restoration of landfills and root penetration**   178
William Young
Response: Horticultural research at the interface of remediation, waste management and site design   *191*

14 **Crissy Field: tidal marsh restoration and form**   193
Kirt Rieder
Response: Natural processes, cultural processes and manufactured sites   *206*

15 **The Sydney Olympics 2000: combining technology and design in the planning of the "Green Games"**   208
Michael Horne
Response: Environmental site design as public infrastructure?   *219*

16 **Sydney Olympics 2000: Northern Water Feature**   221
Kevin Conger
Response: From mechanical to the biological   *239*

PART IV Postscript   241

17 **Dialogues with the contributors**   243
Niall Kirkwood

Figure credits   *252*
Bibliography   *253*
Index   *254*

Contributors

Deborah DiVerdi Carlson is a partner in the business department of Posternak, Blankstein & Lund, L.L.P., Boston, MA and has been practicing law for fourteen years. Ms Carlson has extensive experience in the formation of, and operational issues of concern to, corporations, partnerships, limited liability companies, and other entities. She also handles matters involving federal and state taxation for individuals and business entities on tax-exempt qualification issues and general transaction matters.

Eric Carman is a Vice-President, Senior Scientist and Hydrogeologist with ARCADIS, Geraghty & Miller Environmental and Infrastructure of Milwaukee, Wisconsin and has extensive experience in the use of innovative remedial technologies, and in the management and implementation of large-scale projects involving surface water and soil contamination. Mr Carman serves as the Wisconsin Hub Manager for CERCLA Services and as the coordinator for projects involving phytoremediation and bioremediation.

Kevin Conger is a landscape architect, founder of Conger Design, and an Associate of Hargreaves Associates, Landscape Architecture, Urban Design and Planning in San Francisco, California. Mr Conger was a member of the Hargreaves Associates design team which was involved with the preparation of a new Master Concept Design for the Public Domain of the site for the 2000 Olympics at Homebush Bay in Sydney, Australia.

John S. Day is an attorney in the litigation department of Posternak, Blankstein & Lund, L.L.P., Boston, MA. and represents a diverse range of clients in a variety of contractual and tort litigation matters, including environmental, insurance coverage, products liability, construction and general business disputes. Mr Day advises and counsels clients regarding various environmental law issues, including potential liability under M.G.L. chapter 21E compliance requirements as set forth in the Massachusetts Contingency Plan, demands for reimbursement or contribution of costs associated with cleaning up contaminated property, and potential environmental matters that might arise during the course of commercial property transactions.

Wendi Goldsmith is a geomorphologist and soil scientist with The Bioengineering Group, Inc. She has been involved with planning, design, implementation, monitoring, and research related to stream stabilization, wetland creation, and water quality management. She also serves on the Board of the Center for Urban Watershed Renewal. Ms Goldsmith received her Bachelors in Geology and Environmental Studies from Yale University, her Masters in Landscape Architecture from the Conway School, and her Masters in Plant and Soil Science from UMass Amherst.

George Hargreaves is the Peter Louis Hornbeck Professor in Practice of Landscape Architecture and is the Chairman of the Department of Landscape Architecture at the Harvard University Graduate School of Design where he has taught since 1986. He teaches advanced and theoretical design studios and leads a landform workshop in core studios. He is the principal in the San Francisco and Cambridge firm Hargreaves Associates which has designed a variety of award-winning parks, plazas, and riverfronts. His work has been published nationally and internationally and has won numerous design awards.

Michael Horne is a Landscape Architect and Urban Designer, and was Design Manager of the Government Architect Design Directorate (GADD), which supported the Olympic Coordination Authority in the design of the public domain at Homebush Bay. Mr Horne's project experience extends across both the public and private sectors, including Manager and Landscape Designer at the New South Wales Department of Public Works and Services, and as an Associate at Woods Bagot, one of Australia's foremost multidisciplinary design practices.

Michael Hough is a Professor in the Faculty of Environmental Studies at York University in Toronto where he has taught since 1971. In addition he is a principal and founding partner in the landscape architectural firm of Hough Woodland Naylor Dance Leinster in Toronto, Canada. His firm is recognized for its ecological planning and design work and he has conducted extensive applied research in ecological restoration.

Dr Lucinda Jackson is a Staff Environmental Scientist with Chevron Corporation in Richmond, California. She consults for Chevron operations worldwide on environmental biology issues, including environmental impact analysis, ecological risk assessment, phytoremediation, environmental public affairs, and habitat protection and restoration. She has worked in agricultural and environmental biology since 1973 for several institutions, including the Agricultural Extension Service, FMC Corporation, Monsanto Company and Dow Chemical Company. She sits on the Board of non-profit environmental groups involved in environmental education and habitat preservation.

Niall Kirkwood is Director of Master in Landscape Architecture Degree Programs and Associate Professor of Landscape Architecture at the Graduate School of Design, Harvard University. He previously worked in private landscape practice on the redevelopment of municipal landfills, urban docklands and railroad yards. His current teaching and research addresses concerns in urban brownfields, land reclamation and the alliance of innovative site technologies and progressive site design and planning in contaminated landscapes. He was recently appointed Director of the Center for Technology and Environment (CTE) at the Graduate School of Design, a research, teaching and outreach organization focusing on the reclaimed landscape.

Rebecca Krinke is an Assistant Professor of Landscape Architecture at the University of Minnesota, teaching design studios and courses on emerging topics in landscape architecture. She has also taught studios at the Harvard Graduate School of Design and Rhode Island School of Design. Krinke has worked with landscape architectural firms focusing on the design of urban public space. Educated as both an artist and landscape architect, Krinke's research interests explore the relationships between art, science and innovative design practice. She co-curated the "Manufactured Sites" exhibition.

Deborah Lange is the Executive Director of the Brownfields Center, a joint venture at Carnegie Mellon University and the University of Pittsburgh. She is also a PhD candidate in the Department of Civil and Environmental Engineering at Carnegie Mellon University. Deborah has more than fifteen years of professional consulting experience and has practiced as an environmental engineer in the North, Central and South Americas, and Europe. She is active in professional engineering and environmental associations.

Professor Peter Latz is a founding partner of Latz + Partner, Duisburg, Germany and Head of the Department of Landscape Architecture and Planning at the Technical University of Munich. The range of professional work is concerned with projects of urban planning and design as well as large-scale landscape architecture. These include the planning of open space and ecological construction to research work in the field of alternative technologies. Projects include Potsdam, Peoples Park on the Bornstedter Feld, and Bundesgartenschau, 2001, Landscape Park Duisburg-Nord, and the Revitalization of Böhlen-Lippendorf, Germany.

David Li is an environmental attorney and a partner in the Environmental Law Group of Posternak, Blankstein & Lund, L.L.P., of Boston, Massachusetts. Mr Li represents corporate and municipal clients in litigation and regulatory compliance matters arising from federal and state environmental laws, including Superfund, RCRA, the Clean Water and Clean Air Acts as well as Massachusetts Chapter 21 E. Prior to his legal career in the private sector, Mr Li served both Maryland and Massachusetts as an assistant attorney-general.

Deborah Marton is a project manager and landscape restoration designer for the Natural Resources Group, New York City Department of Parks & Recreation. Ms Marton's current projects include designs for restoration of a kettle pond with wildlife viewing area, excavation of fill and replanting of a salt marsh, and co-design of an ecology park and maritime shrubland restoration at Paerdegat basin in Brooklyn. Ms Marton was formerly principal of her own consulting firm, specializing in landscape restoration design, particularly in the area of landfill restoration, and relevant legal compliance. Prior to consulting, she worked as a corporate litigator at Wilkie, Farr and Gallagher in New York, and is a member of the New York Bar.

Dr Sue McNeil is Director of the Urban Transportation Center and Professor of Urban Planning at the University of Illinois, Chicago. Her research and teaching interests focus on infrastructure management with emphasis on the application of advanced technologies, economic analysis, analytical methods, and computer applications. Dr McNeil initiated and chaired the ASCE Urban Transportation Division Committee on Transportation Facilities Management (1988–93). She is an Associate Editor for the ASCE *Journal of Infrastructure Systems*, and a former director of the Brownfields Center at Carnegie Mellon University and the University of Pittsburgh.

Richard Owen is an Associate Director of Arup Environmental and a Chartered Civil Engineer with twenty-five years' specialist consulting experience in the fields of Geotechnical and Environmental Engineering. Over the past twelve years Mr Owen has been principally involved in contaminated land and environmental assessment work. He has directed many projects and provided expert advice in the areas of soil and groundwater contamination, waste disposal, landfill gas emissions, air quality, environmental auditing and assessment. Mr Owen is involved in the development and production of Contaminated Land Guidance and Policy. In particular he is an active member of BSI committees updating the Codes of Practice for Site Investigation.

Kirt Rieder is a landscape architect and Associate of Hargreaves Associates, Landscape Architecture, Urban Design and Planning in Cambridge, Massachusetts. Mr Rieder is a member of the design team involved with the Crissy Field Project located at the Presidio in San Francisco. In addition, project work includes Stapleton Airport I-70 Frontage Landscape Concept Design, Denver, Colorado and Parque do Tejoe Trancão, Expo '98 in Lisbon, Portugal.

Dr Steven Rock is an environmental engineer with the National Risk Management Research Laboratory of the US Environmental Protection Agency in Cincinnati, Ohio. Dr Rock currently works in the Remediation and Containment branch conducting and coordinating field research into phytoremediation and is co-chair of the Phytoremediation Research Technology Development Forum (RTDF). He has also worked in the Superfund Innovative Technology Evaluation Program and has authored numerous publications in the field.

Dr Jean Rogers is Director of Strategy for Razorfish – San Francisco, a firm that takes clients into the digital realm in every possible way, including web and wireless solutions for their operations, products and services. She currently designs high-tech, low-impact solutions for clients ranging from Fortune100 manufacturing concerns to a start-up that is designing a website to teach kids about science and technology. Her background is in engineering, design planning and management consulting. She was a Loeb Fellow at Harvard Graduate School of Design in 1997/8, where she taught and did research on achieving competitiveness through design. She adheres to the philosophy that everything that can be digital will be, thereby preventing many of the remediation scenarios encountered in this book.

Rosanna Sattler is an environmental attorney and a partner and chair of the Environmental Group of Posternak, Blankstein & Lund, L.L.P., of Boston, Massachusetts. Ms Sattler has earned a national reputation for her experience in environmental compliance and litigation. She represents and advises businesses, environmental service companies, engineers, consultants, property owners, municipalities and other entities in disputes with regard to regulatory compliance issues. Ms Sattler works closely with hydrogeologists, engineers, toxicologists, chemists and site remediation consultants in connection with site assessments, studies and risk assessments. Ms Sattler has written and lectured extensively on legal issues affecting the environmental industry and is a member of the American Bar Association Section of the Natural Resources Energy and Environmental Law, the American Bar Association Environmental Litigation Committee, the Boston Bar Association Environmental Law Section, the Massachusetts Academy of Trial Attorneys and Boston Inn of Court.

Dr Dante Tedaldi is an environmental engineer and a technical and management consultant with Bechtel National, Inc. of San Diego, California, specializing in strategic planning, policy formulation, environmental management systems and program implementation. His experience includes operations review, feasibility studies, and implementation of high-impact projects including Project Engineer for the Asarco Smelter Superfund site, the Department of Navy Southwest Environmental Program, and work for the US Department of Energy, General Electric Corporation and FMC Corporation.

Lorna Walker is a director of Ove Arup & Partners and leader of Arup Environmental. As a chartered chemist and qualified civil engineer of twenty-five years' experience, Ms Walker is a renowned expert in the fields of water quality and water and waste treatment. A member of a number of policy-making panels in the industry, she is frequently asked to assist in the preparation of proofs of evidence for planning enquiries and as an expert witness in litigation cases. As Project Director, experience includes chemical and biological investigation of polluted dock water and the design and implementation at Salford Quays, Manchester, UK and environmental project work at Potsdam Plaza, Berlin, Germany, and Sagrera Development, Barcelona, Spain.

Daniel Winterbottom is Assistant Professor of Landscape Architecture in the Department of Landscape Architecture, University of Washington, Seattle. His research and teaching focus on the relationship of landscape technologies to site design and the application of sustainable and local technologies in the development of community resources.

William Young is a landscape architect, wetland specialist and Director of Restoration Projects for The Dawson Corporation of Clarksburg, New Jersey. His areas of expertise include habitat restoration on disturbed lands, wetland monitoring and construction, bio-engineering of streams and rivers and erosion and sediment control design. Project experience includes conversion of sludge ponds to grassland habitat for Bayer Corporation, South Carolina, and the restoration and end-use plan, wetlands mitigation, and permitting, designing and construction of stormwater retention basins for Fresh Kills Landfill, Staten Island, New York. Mr Young was formerly a senior scientist with PBS & J. in Florida.

Foreword

The way we design our cities has always, sometimes unwittingly, been determined by what history, societal values, and the fundamental natural laws of change and adaptation have left us. Already in evidence are the emerging efforts to address the legacy of contaminated and derelict lands that have been left by past industrial activity. They are everywhere in evidence: along urban waterfronts, in the wastelands associated with urban rivers, and in outlying areas – those places that have become known as manufactured sites. These little used, economically and environmentally degraded landscapes, often contaminated in varying degrees with a noxious mix of chemicals, pose major challenges at every level: for soil and groundwater sciences, engineering, legal and regulatory agencies, for the politicians, for planning and design, and for neighborhood communities and the public at large. It is, in effect, society at large that must determine their future uses, and how best to restore and integrate them into the urban fabric.

While the recycling of derelict industrial land to alternative urban uses has been accepted practice for many years, it has frequently been accompanied by developments that have too often done little to enrich the environments they have replaced. Those who attended waterfront renewal conferences in the 1970s may be forgiven for wondering why the "before" photos of the site so frequently looked more interesting and alive than the built form that emerged. The era of the large-scale master plan has taken its toll in many waterfront cities. There is no doubt that the challenges today have become increasingly complex the more that is known about these sites. At the same time, there are enormous opportunities for new thinking that is appropriate to a new and progressive century where the impetus of societal concerns is moving toward a philosophy of sustainability.

From long personal and professional experience of my hometown, there is no doubt that Toronto, like most other North American cities, has had its fair share of degraded and economically unproductive landscapes, especially on its waterfront. It illustrates the essential lessons that are already being understood: that they are an immensely valuable resource for the city's changing future and

- that there is an emerging trend, away from the exclusive domain of engineering, or any other discipline for that matter, to a search for solutions that are becoming increasingly collaborative and integrative;
- that these places need environmental, economic, legal, policy, and site and urban design expertise if they are to be restored to environmental health and integrated into the economic and social life of the city.

Other factors will be also be influential in the way in which we approach the renewal of old industrial areas. Of particular significance is the conceptual framework that is challenging established conventions: that economics and urban development are the determining priorities for renewal, leaving other environmental, ecological and social factors as poor cousins, with little voice in the redevelopment process. Recent initiatives in North America and Europe are revealing a different kind of thinking at work. The restructuring of old industrial areas in Germany has begun with ecological, *not* economic, renewal as recent experience in the Ruhr Valley is demonstrating. Proposals to introduce networks of multi-functional corridors and parks in Toronto's Port Lands is seen as an essential environmental framework for future development – establishing landscape as the precursor to built form. In Sudbury, Ontario, the restoration of forest from lands sterilized by past nickel processing also illustrates how new approaches to rectifying past practices are taking hold.

These abandoned sites are often ecologically rich, with regenerating vegetation and habitat that somehow manages to thrive on frequently hostile soils. Industrial artifacts may provide clues to their history and inherent sense of place. Concerned and activist local communities are already having an impact on the politics of urban renewal, a trend that will factor into the political future of such lands. There is also the realization across North America that uncontrolled expansion of urban growth has been accompanied by enormous consequences for the environment. These include energy use, the impact of air pollution on human health, and quality of life, at a time when efforts to reach the goals of sustainability are becoming a priority at local, national, and international levels.

This book, *Manufactured Sites*, is a landmark in current and progressive approaches to the issues facing such sites. It takes a major step in examining and providing a wide range of practical, implemented solutions, such as phytoremediation about which there is much discussion but little application. Of central importance is how an integrated approach to restoration provides the essential opportunities for dealing with the many urban problems that have faced society in the past, and will undoubtedly continue to do so. These industrial places have been called the parks of the twenty-first century. It's a vision that is entirely appropriate to the future of our cities and the direction that must be taken. This timely and urgently needed book points the way.

Michael Hough
Principal and Founding Partner,
Hough Woodland Naylor Dance Leinster
Professor, Faculty of Environmental Studies,
York University, Toronto, Canada

Preface

Manufactured Sites concerns the reclamation of waste and contaminated urban sites. It presents scientific research, environmental construction and creative landscape design practices that support the cleaning up and regeneration of derelict and toxic urban land and waterways.

Over the last twenty years there has been a need to address the damaging legacy of derelict and abandoned industrial buildings, environmentally compromised land areas and polluted soils and water. Federal initiatives and legislation on the clean-up of these urban sites, their critical location within regional transportation centers, and the diminishing number of "greenfield" sites available for development have all acted to focus research, engineering and design efforts on their redevelopment and reuse. It has also been recognized that environmental problems encountered on these contaminated sites are sufficiently complex that they require scientists, engineers, designers, and planners (among others) to work collaboratively across disciplines and institutional boundaries to address major challenges in remediating and planning urban land.

A focus throughout this book is not whether these sites should be reclaimed, restored or redeveloped, but, rather, the precise nature of how this is carried out, and the opportunities brought by these collaborative and integrative exchanges. Therefore, two central themes of the book are the range of emerging technologies and design strategies used in reclaiming waste and contaminated urban sites and the creative alliances of technology and design that result.

Manufactured Sites comprises a series of essays prepared for this publication by an international group of academics, engineers and designers. Responses that accompany each of the essays discuss approaches taken by each of the contributors to the integration of scientific, environmental, engineering and design information and concerns, as well as commentary on the results of built projects.

Intended primarily for landscape architects, planners and environmental practitioners, *Manufactured Sites* introduces advances in specialized environmental technologies used in the clean-up and recycling of contaminated sites alongside progressive construction technologies and design practices used in the recovery of land altered by industry. It is my intention however that other design practitioners, academics and students will recognize in the essays significant opportunities within the alliances of technology and design to influence their own creative work and undertake new professional and research initiatives.

Origins of the book

In the spring of 1998 a conference and exhibition organized by the author, and sponsored by the Department of Landscape Architecture, took place at Harvard University's Graduate School of Design. Entitled, "Manufactured Sites, A Landscape Conference on Site Technologies for Contemporary Practice" it featured presentations over two days from invited speakers from Australia, Germany, England and the United States. An audience of over two hundred practitioners and students, from a broad range of engineering and design fields, also participated in question and answer periods and break-out discussion sessions. New research and case studies of innovative site technologies and their applications to a range of manufactured sites were featured. Topical professional and industry projects were highlighted – for example, waterfronts and wetlands, large-scale landscape infrastructure, and waste and water management proposals.

This book brings those speakers together again and revisits a number of the projects that were presented at the conference. Each speaker was asked to prepare edited versions of their presentations and to further reflect on their subject matter as well as the work of other presenters. I am happy to say that most agreed it was a worthwhile thing to do. In addition, the conference moderators have written responses to each of the essays and these broadly dwell on individual interpretations of the interdisciplinary themes raised at the conference.

Acknowledgements

A number of other individuals have provided commentary and critique throughout the period from conference to manuscript. These include Alistair McIntosh, Partner of Sasaki Associates, Landscape Architects, Watertown, Mass., Julie Bargman, Assistant Professor in the Department of Landscape Architecture, University of Virginia, and Dr Susanne Hauser of the University of the Arts, Berlin, Germany. I am grateful for their insights. Peter Rowe, Dean of the Faculties of the Harvard Graduate School of Design is to be acknowledged for his continued support in the development of this area of academic research, as is Professor George Hargreaves, Chairman of the Department of Landscape Architecture, for the Department's continued encouragement, assistance and focus on this emerging class of work. All the writers are to thanked for their contributions to the book. They are, in alphabetical order, Deborah DiVerdi Carlson Eric Carman, Kevin Conger, John Day, Wendi Goldsmith, Mike Horne, Lucinda Jackson, Deborah Lange, Peter Latz with Anneliese Latz, David Li, Sue McNeil, Richard Owen, Rosanna Sattler, Kirt Rieder, Steve Rock, Jean Rogers, Dante Tedaldi, Lorna Walker and Bill Young.

Michael Hough is also to be thanked for writing the foreword to this book, as is Deborah Marton for her earlier work as a moderator during the conference.

A large amount of work accomplished toward this publication was carried out by two people: Rebecca Krinke, now Assistant Professor in the Department of Landscape Architecture, University of Minnesota, and Daniel Winterbottom, Assistant Professor in the Department of Landscape Architecture, University of Washington. I wish to acknowledge their immense contribution in preparing essays and responses and taking the time away from their own busy design practices and research.

An exhibition, also titled "Manufactured Sites," was held in April 1998 at the same time as the conference. This was organized and curated by Niall G. Kirkwood, Associate Professor of Landscape Architecture, Rebecca Krinke, Visiting Critic in Landscape Architecture and Brooke Hodge, Director of Exhibitions, with support from George Hargreaves, Chairman of the Department of Landscape Architecture, Harvard University Graduate School of Design. Final assistance in preparing for the exhibition was provided by Kim Everett and James Stone of the Graduate School of Design. I want to acknowledge the following lenders for the exhibition and the individuals in the organizations and design practices without whose efforts in researching and providing documentation material it would not have been possible: Julian Hart of Arup Environmental, Ove Arup and Partners, London, England, including copyright approvals from John Donat and Marcus Taylor Photography. Clarrisa Rowe and Lisa Roth of Brown and Rowe, Inc., Boston, Mass. and The Central Artery/Tunnel Project, Boston. Kirt Rieder, Kevin Conger and Willet Moss of Hargreaves Associates, Inc., San Francisco, Calif. Michael Horne, Government Architectural Design Directorate, Sydney, Australia. Anneliese Latz of Latz + Partner, Kranzberg, Germany. Deborah Marton of the Natural Resources Group, New York City Parks & Recreation. Martha Schwartz, Lital Szmuk and Patricia Bales of Martha Schwartz Inc. Cambridge, Mass.

I wish to thank Rebecca Casey, Associate Editor of Spon Press, London for her guidance during the preparation of the manuscript and publication period. In addition Chris Matthews of the Harvard Graduate School of Design for his assistance in assembling and preparing a number of the images featured in the essays, and Efthimia Paikos, the administrator of the landscape Program Office for her help with the preparation of the manuscript.

Finally I wish to acknowledge the interest and enthusiasm shown by my students at the Harvard Graduate School of Design toward the subject of manufactured sites. These are the young professionals who, in their future work, will inherit the sites that are under discussion. It is reassuring to find them already deeply engaged in both the subject matter and the design possibilities that will arise from manufactured sites.

Niall Kirkwood
Cambridge, Massachusetts
June, 2000

Part I
Introduction

Part I introduces manufactured sites and provides an overview to the reclamation of waste and contaminated urban sites. It also serves as an introduction to the subject matter of the essays and responses that follow in Parts II and III.

In the first essay of Part I, an overview of manufactured sites is presented, along with recent academic, professional and industry concerns in this area. Three definitions of the term "manufactured sites" are introduced that address the conditions of these contaminated urban sites. Two central themes are introduced: the range of emerging technologies and design strategies used in reclaiming waste and contaminated urban sites, and the promising and creative alliances of technology and design that result.

Rosanna Sattler and David Li, experienced and nationally respected environmental lawyers, have been closely involved in the development and review of legislation and litigation surrounding manufactured sites and environmental contamination. In the second essay, assisted by their colleagues Deborah DiVerdi Carlson and John S. Day, they lay out the legal and regulatory concerns of manufactured sites. These concentrate on environmental law and the magnitude and uncertainty of environmental liability risks and costs. Although focused primarily on the United States, mention is also made of similar laws in Europe and the United Kingdom. Sattler and Li present and reinforce the role of environmental lawyers as playing prominent and active roles as members of multidisciplinary environmental and design teams.

Chapter 1

Manufactured sites:

integrating technology and design in reclaimed landscapes

Niall Kirkwood

> Virtually every city in the nation's older industrial regions, no matter its size, grapples with the challenge of unused manufacturing facilities and other industrial sites ... Public concern about health effects from hazardous chemicals, stricter environmental laws, and changing private-sector development priorities have made it increasingly difficult for communities to restore and reuse former manufacturing sites.[1]

Introduction

What are "manufactured sites"? As Michael Hough states in the Foreword, manufactured sites are related to "emerging efforts to address the legacy of contaminated and derelict lands that have been left by past industrial activity."

The purpose of the collection of essays in this book is broadly to tackle this question from a range of academic, research and practice viewpoints. The invited authors were selected to expose the reader to innovative environmental, engineering and design approaches to the legacy of contaminated and derelict land, as well as to illustrate ongoing research and built projects of national and international significance. In addition, *Manufactured Sites* is seen by the editor and authors as contributing to an ongoing dialogue on the interdisciplinary nature of this work that must occur between those disciplines and professionals who are currently engaged on the restoration and reuse of industrial land in our urban landscape.

The phenomena of manufactured sites that follows is focused on five interrelated areas that will appear with varying emphasis in the essays. First, industrial lands, ranging from small factories to vast municipal landfills that are currently undergoing reuse are described. Second, the range of concepts, and implementation strategies of site technologies that clean up contamination on these sites is illustrated. Third, the professional nature of the recycling of industrial urban land is discussed, addressing attitudes to restoration and uses that are often in conflict with current development and land ownership patterns. Fourth, the types of professional practitioners, industries and specialists involved in the remediation and regeneration of manufactured sites are addressed. Finally, models for reclaiming landscapes are explored based on the alliance and interaction of clean-up technologies with progressive landscape design practices.

3

Where are manufactured sites located?

A derelict three-acre site located next to the Detroit riverfront with locked gates, abandoned machinery and manufacturing wastes and chemical pollutants in the adjacent ground is illustrated in Figure 1.1.

Thousands of manufactured sites like this are now to be found in other major metropolitan centers. They include factories, scrap-metal yards and railroad corridors, waste transfer stations, vacant corner gas stations and rusting machine shops – a mere listing of these places does little to describe the extent of this industrial legacy nationwide.[2] Nor do the sites themselves suggest any single pattern for their future regeneration – they appear initially too haphazard in their nature, size, location and extent of environmental degradation. The pressures for redevelopment are immense, the factors of ownership and contamination are significant, and the visions for a site's reuse are as numerous as there are interested parties.

"Land long polluted"

Reporting on manufactured sites is topical. A 1998 front-page article in *The New York Times*, subtitled "New Laws and Funds Fuel Revival of Land Long Polluted"[3] reported on an emerging industry of financial consultants and environmental scientists that has grown up to attend to the approximately 6,500 "brownfields" in the former manufacturing centers surrounding the city of New York. "Brownfields" are defined here as "abandoned or underused industrial and commercial sites where redevelopment is complicated by real or perceived contamination,"[4] but the term is more commonly associated with land that has immediate potential for economic redevelopment. These sites account for "more than

Figure 1.1 Manufactured site, Detroit riverfront, 1999

1,300 acres of pollution-stained empty lots or industrial buildings, making up 40 percent of the total of 3,300 acres of tainted sites in the city." These sites form part of the estimated 500,000 brownfields considered to be present in the United States at this time.[5] The article also noted that "The land is desirable for the reasons businesses once favored inner city locations – proximity to transportation and easy access for suppliers and customers." However, it concludes that "No one involved in rehabilitating old urban sites expects all the scars of a century of heavy industry to be erased." The continued presence of on-site industrial and chemical wastes, metals, solvents, underground oil tanks and toxic sediments left on-site from industrial activities, renders them problematic for conventional redevelopment strategies in comparison to non-contaminated urban land or rural "greenfield" sites.

In 1998, *The New York Times* again reported a related topic: the fate of the 3,000 acre Fresh Kills Landfill, the municipal waste site for the New York Boroughs on Staten Island.[6] More than three and a half times the size of New York's Central Park, this landfill is slated to be closed by the end of the year 2001 at an estimated cost of $1 billion dollars. Encapsulation and an array of landfill technologies to clean venting gases and to collect leachates were outlined, but the article went on to conclude that "detailed plans have still to be worked out for what to do with the Fresh Kills land." The City of New York, Department of City Planning (DCP) is now considering a number of paths to provide for end-use proposals, including the mounting of public design competitions. Third, a new natural *in situ* remediation technology using plants to uptake heavy metals found in contaminated residential soils in Dorchester, Massachusetts was headlined in *The Boston Globe* in March of 1997 under the title "Plants doing the dirty work in cleanup of toxic waste."[7] The lead contamination in the soils in this case resulted from industrial processes formerly located on an adjacent site.

These three press reports, representative of the growing number throughout the country, draw attention to the derelict and hazardous state of post-industrial land within cities and towns. In addition they record an ever-increasing tendency toward both the need to clean up and redevelop environmentally compromised land.

Reclaiming and recycling

The subject of reclaiming and recycling manufactured sites exerts a special fascination that is not fully explained by the influence it has had on the planning and rebuilding of cities. Even those unconcerned with the daily work of planning, site design and real estate development are now required to take account of this phenomenon. Journalists write about the pressing civic burden of closed landfills, polluted waterways and wetlands, industrial oil spills and abandoned and rusting factory sites. Politicians have taken up the cause of "contaminated urban land" through the growing amount of federal and local state legislation. Community pressure groups were present and vocal from the very beginning. Many groups, whether social, cultural, commercial or political in nature, have laid claim to this disputed area.

Two issues dominate current discussion of these blighted landscapes. First, the remaining pervasive pattern of hazardous substances found in the soils, groundwater and within the fabric of buildings and infrastructure left standing on-site. Second, the motivation to return these manufactured sites to productive use and the physical means by which this can be carried out.

Defining manufactured sites

The term, "manufactured sites" and the title of this book can be understood in three ways. The first refers to a class of site found in older manufacturing cities and towns and whose present condition is a result of manufacturing and industrial processes or disposal of waste. In addition these conditions have now become their identifying and prevailing character. Second, it can signify both the presence of environmentally challenged sites and the relationship to processes and techniques used to clean up these conditions. Third, it presents an interdisciplinary approach to reclaiming sites altered by industrial activity.

"Sites of manufacture"

The first definition, "sites of manufacture," describes sites whose former activities have included industrial and manufacturing processes. Places synonymous with this environmentally compromised condition are easily recognizable within the blighted city fabric: abandoned mills alongside river corridors and canals, mountainous waste landfills that occupy the city's periphery, chemical and oil refinery facilities, derelict marine terminals, and the patchwork of small factories within older residential communities. Examples of "sites of manufacture" include former tool and die shops, metal fabrication and electroplating factories, textile and paper mills, and former manufactured town gas plants that are found in abundance throughout older cities.

Figure 1.2 illustrates the site of a former manufactured town gas plant in a Connecticut city. The circular outline on the ground in the middle of the image denotes the location of the former cylindrical structure of the gas tank holder. Coal tars, including benzene and phenols, as well as metals and cyanides from the town gas manufacturing process, are likely to be found in the surrounding soils and groundwater. In this case, the term, "sites of manufacture" is used to identify and classify land – particularly if, as is often the case, it occupies prime locations within the urban fabric and is currently sought after for economic redevelopment. In addition the term also applies to less economically desirable land for development that still supports neighborhoods in terms of providing future sites for community amenities.

Figure 1.2 Former gas manufacturing plant, Connecticut, 2000

Manufacture of sites

The second definition, "manufacture of sites," is concerned with the environmentally compromised nature of these sites that has resulted from earlier periods of manufacturing activity. Often invisible to the naked eye, these environmental problems involve the continued presence of metals, oils and chemical residues of industrial processes, as well as the uncontrolled illegal dumping of waste on the site that continues. The term applies not only to these former locations of industrial activity and their toxic by-products but also to the application of on- and off-site environmental construction and engineering technologies that systematically clean up and "manufacture" the materials of the site, whether soils, groundwater or infrastructure elements. Examples of these operations include the movement, dispersal, encapsulation or treatment of soil and fill material as well as the use of engineering and biological systems to treat and reshape earth and groundwater flows and to stabilize ground surfaces and landforms.

Manufactured sites as a integrated redevelopment process

Finally, "manufactured sites" can also be viewed as an approach to the regeneration of contaminated urban land. This approach looks to the dynamic nature of the existing physical site conditions, however degraded, and the integration and alliances of clean-up and engineering technologies with design strategies to guide future redevelopment.

Just as traces of a site's past are developed into design proposals, the logic and needs of a technology can inspire and shape thinking about conceptual site design. Spatial, experiential and aesthetic clues can be drawn from technologies and these in turn can begin to inform design proposals. In short, site technologies are not viewed as physical constraints but as means of inspiration.

In manufactured sites during the process of integrated redevelopment, designers look to the spatial and systematic form of remediation technologies introduced onto the site and the interaction of these technologies with progressive landscape design practices in the systematic production or "manufacture" of future sites. Project proposals and redevelopment strategies are no longer seen as the means to impose further changes or enlargements to the site area, rather they are integrated into a more holistic view of contaminated land reuse.

"Parks not poisons"

From the politicians and lawmakers who have sought to regulate the clean up of these sites, to the communities, developers, lawyers, planners, engineers, designers who work to regenerate them and reuse them, environmentally disturbed sites and their reclamation have become a focus of private–public initiatives and partnerships. The significance of manufactured sites has been underscored at the federal level. In the 1997 State of the Union Address, President Clinton stated that: "We should restore contaminated urban land and buildings to productive use . . . In last four years, we cleaned up 250 toxic waste sites, as many as in the previous twelve [years]. Now, we should clean-up 500 more, so that our children grow up next to parks, not poison."[8]

Federal attention continues to be given to waste sites through initiatives and legislation. For example, only recently a new reform bill – the Brownfields Revitalization and Environmental Restoration Act of 2000, was announced by a bipartisan coalition in June of 2000.

Others have disagreed, however, pointing out the insignificance of these manufactured sites in relationship to more pressing concerns in the environment such as the continuing effects of over-development, particularly sprawl. These commentators and critics ignore the fact that recycling and reuse of these sites is an old activity, and one that is likely to continue to influence sprawl through the rebuilding of the urban core and reducing further encroachments into "greenfield" sites. The need to revive the physical, economic and environmental health of cities and urban communities has led to the development of federal and local state laws and initiatives aimed at promoting the reclamation of derelict land. The purposes of these policies are to return polluted land to economic benefit and uses and to promote healthier and more sustainable urban environments.

Civilization's toxic mixture

Civilization's mix of former manufacturing and waste sites varies in size and intensity of contamination. Scales of sites vary from the 3,000-acre Fresh Kills Landfill noted in *The New York Times* article and discussed by William Young in Chapter 13, to the single former gas station corner lot found within a residential neighborhood. Environmentally damaged sites such as these are under a growing scrutiny from interested parties and stakeholders, including federal and state agencies, local municipalities and landowners, scientists and corporations, environmental groups and developers, community groups and utility companies, biologists and engineers. Manufactured sites have also gained broader recognition by public and private development agencies, through their prime location within existing infrastructure, their key locations within regional and national transportation systems and their adjacency to existing city centers with local resources of labor and supporting services. Acknowledged redevelopment "problems" are attached to them, however, arising from issues of regulatory compliance, ongoing liability relief and required standards of clean-up. These will be addressed by Rosanna Sattler and David Li in Chapter 2.

Professional groups and professional work

There are two groups of professional consultants and researchers who have productively but independently worked within land reclamation and reuse: environmental and civil engineers and the growing number of site designers, whether landscape architects, site architects or urban planners and designers. A large body of scientific research into emerging and innovative environmental technologies has been carried out, and Jackson in Chapter 3 and Rock in Chapter 4 describe examples. However, most of this work lies beyond the daily review of the design reader, concerned as it is with scientific data on the nature and remediation of toxic materials and the results and analysis of laboratory and field-testing. At the same time, much of the writing about progressive site planning and design approaches for urban regeneration is little read outside of that field.

Lessons from the field

In the fall of 1998, a number of graduate landscape design students and I stood at the highest plateau of a former municipal landfill in the northern part of Cambridge, Massachusetts at what is now called Danehy Park. From this vantage point we could see an open green public park, 60 acres in size, comprised of grassed slopes and meandering pedestrian pathways, soccer fields and baseball diamonds, a busy children's playground, joggers and cyclists and a thriving 2-acre wetland teeming with birds, insects and aquatic plants. The use of the land has changed over time from use as a brick pit in the early nineteenth century, to a municipal landfill and spoil dump, then to the present park.

Walking further, the group discovered methane gas gravel trenches around the site, citron and brown tinted pools in the lower grassed areas next to the wetland, monitoring wells – all elements not commonly associated with the public parks and open space. The group also looked very closely at the ground. Drainage was poor, but not unduly so; the absence of shade trees was noticeable for a park, but not significantly so; the contouring of the ground, however, indicated an unusual topographical condition. The hilltop plateau and the surrounding land were shaped in squat and angular forms, constructed as a process of systematic filling and capping. The previous landfilling operations, the "lifts" of garbage and waste, could be seen in relation to the current grassed slope profiles.

Here, at the interface between terra firma and the sub-ground, as illustrated in Figure 1.3, was found the connection between technologies employed to store and contain (or cap) the landfill material and the final site design proposal – a relationship based on the morphology of a site "manufactured" over time, and the structure and form of a public park that echoed this morphology. As for the waste it had not disappeared, it was simply out of sight, with constant reminders through the presence of venting trenches and monitoring wells.

Unlike past innovations in landscape design and city planning, which responded to economic change, or growing alterations to population or technological invention, this type of design work is a return to the productive use of exhausted and currently undervalued plots of ground – a tidying up of the past industrial environment. These landscapes are initially dependent for their reuse on the application of a wide range of site engineering and environmental reclamation technologies. These include landfill capping, on-site techniques of soil and groundwater clean-up, bioengineering, the alliance of biological and engineered systems used in the remediation of site contamination, wastewater management systems, and environmental monitoring. One of the tasks to be undertaken by scientists, engineers and site designers together is to find innovative ways to unite these activities with new site programs and uses.

Figure 1.3 Examination of trial inspection pit revealing thin soil layer and landfill cap, Danehy Park, Cambridge, Massachusetts

9

Integrating remediation and landscape design

The following describes a possible site proposal following such an interdisciplinary exchange. Eight lines of 200 poplar trees interspersed with other non-phytoremediation species are located at the entrance to a new public open space – a community park, the interim use of a manufactured site. The poplar trees are geometrically spaced according to their requirements to naturally intercept and uptake subsurface groundwater pollution that is moving off-site. Throughout the open space small fenced patches of contamination "hot-spots" are treated behind screens of densely planted trees and thick grasses, while larger capped areas contain recreational fields and open areas of grass and pathways. In time, this may become the location of new building programs (for education, industry, housing, or community facilities). This hypothetical example of the integration between science, engineering, remediation, and design and planning illustrates a regenerative approach to manufactured sites that uses the application of a natural remediation system. The remediation of the site establishes a physical structure on-site, a vegetation framework of canopy trees, groundcovers and grasses. These in turn equate with the location of future development proposals, including building structures, roads and utilities and open space, or establish a flexible design framework to be acted upon at a later date.

Purpose of this book

The book will review the integration of biological, chemical and engineering sciences with site planning, design and landscape architecture. It will also consider cases of manufactured sites where full-scale remediation is not required, or even desirable. It examines innovative approaches to contaminated sites that bridge the concerns of environmental action and regeneration design. A specific emphasis will be on scientific research, site technologies and landscape design where there is an interrelationship in the remaking and reuse of these neglected sites. The following question will be addressed: "how can current scientific research and technologies support and result in more creative design work?" or, conversely, "what is the landscape design contribution to this area of work?" Of equal importance will be the issues of how will designers and engineers work together in the future, and what will be the nature of their interrelated work?

Structure of book

The book is structured in four parts. Part I provides an overview to the current background and conditions of manufactured sites. This is accompanied by a thorough explanation of the environmental law and regulatory framework that surrounds and influences this class of site. In Part II and Part III, fourteen essays with accompanying responses address the subject of manufactured sites from a range of research, professional, industry and academic viewpoints. In Part IV, discussion of issues and approaches raised in the essays are further described through dialogues with the contributors.

Essays

The breadth and depth of the information contained within the essays is formidable, running as it does from applied scientific research in the field, to site engineering principles, to applied landscape planning and design. Topical professional and industry projects, for example waterfronts and wetlands, large-scale landscape infrastructure, and waste and water management are highlighted, challenging the conventions of current design practice and thinking.

Project case studies range from examples of scientific research and fieldwork in the remediation of contaminated groundwater and soils; others focus on the patterns of environmental management in complex infrastructure and industrial sites; yet others present current progressive design work involving the planning and design of recycled urban land.

Summary

"Manufactured sites" as a model for urban development in the next century creates opportunities for collaborative work between the scientist, engineer, planner and designer. It also provides the means both to protect and improve the natural and man-made civic environment where the recycling of inner city industrial sites relieves pressure on undeveloped land and further encroachment on "greenfield areas."

Finally, it should be noted that this represents another cycle in the reuse and redevelopment of these sites. Land that undergoes radical change today with innovative processes of clean-up and design reuse will themselves be remade in time as the requirements of new forms of program evolve for the workplace, and for living and recreational urban space. Until then "manufactured sites" is presented as a working paradigm for rethinking the post-industrial landscape.

Notes

1   Bartsch, Charles and Collaton, Elizabeth, (1997) *Brownfields. Cleaning and Reusing Contaminated Properties*, Westport, Conn.: Praeger.
2   United States Accounting Office, 1998.
3   Revkin, Andrew C., "For Urban Wastelands, Tomatoes and Other Life – New Laws and Funds Fuel Revival of Land Long Polluted," *New York Times*, Tuesday, March 3, 1998.
4   United States Environmental Protection Agency (USEPA), Definition, 1996.
5   Revkin, Andrew C., *New York Times*, Tuesday, March 3, 1998.
6   Toy, Vivian S., "Worlds Largest Clean-up" *New York Times*, December 1997.
7   *Boston Globe*, March 10, 1997.
8   State of the Union Address, February 4, 1997.

Chapter 2

New designs in the legal landscape

Rosanna Sattler and David Y. Li with
Deborah DiVerdi Carlson and John S. Day

Introduction

Environmental laws that were passed during the latter half of the twentieth century have played a large, if not primary, role in stifling the redevelopment of manufactured sites. These laws impose broad legal and financial liabilities on various individuals and entities associated with manufactured sites, including current owners, potential purchasers, and lenders who provide capital to finance the purchase of such sites. These liabilities create strong disincentives to the redevelopment of urban areas that contain actual or suspected contamination. Developers and prospective purchasers shy away from such properties, despite their often desirable location, on account of the uncertain and seemingly unending liability associated with them. The regulatory risks and liability risks have been created by the laws governing polluted sites. The regulatory risks arise from a myriad of complicated environmental laws that regulate such matters as impact assessment, standard-setting and licensing. Compliance with these laws can be expensive, time-consuming, onerous and, in some cases, may preclude the development of a manufactured site or the operations of a business on the premises. Failure to comply with these laws and regulations can result in excessive costs, as well as exposure to civil and criminal fines and penalties.

Fortunately, the disincentives created by legislation which imposed broad environmental liabilities are in the process of being dismantled by a variety of governmental and private initiatives. These initiatives, which include favorable revisions to the "traditional" environmental laws, innovative insurance options, and tax incentives, share one purpose – the redevelopment of manufactured sites that have been abandoned, undeveloped and/or unused. However, in spite of the active development of these issues, manufactured sites continue to pose significant liability risks to owners, developers, potential purchasers, and lenders. Furthermore, these developing initiatives are most effective if all interested parties collaborate at every stage of the redevelopment process. Accordingly, any group of individuals or entities involved in the redevelopment of a manufactured site should include an environmental lawyer who is well-versed in the various legal requirements and options available for the productive development of such sites.

The magnitude and uncertainty of environmental liability costs create frustrations and obstacles to the redevelopment of contaminated urban sites. Prospective purchasers and developers shun ownership of urban industrial property because they fear liability for past contamination and the costs of complying with the complicated environmental regulatory requirements. Lenders are reluctant to invest in manufactured sites due to prospective lender liability, and the fear that discovery of contamination will erode the value of property as collateral, forcing debtors to default on loans used to develop such sites. Even when developers and lenders are interested in a project, the threat of mandatory assessment and remediation, and the uncontrollable costs associated with these activities, makes the development of budgets, critical paths and schedules difficult.

Overcoming the multiple obstacles to the successful development of a manufactured site can be an arduous process. Nonetheless, developers, manufacturers, businesses, environmental engineers, landscape architects, lawyers, community groups, lenders, and government parties are attempting to remove these barriers to industrial redevelopment. Voluntary clean-up programs and government incentives are emerging as innovative reforms in the area of hazardous waste clean-up and the development of manufactured sites. Governmental regulatory frameworks and public and private programs have been developed specifically to address reduction or eradication of legal liability, development of more flexible technical requirements, and innovative economic incentives. These reforms are aimed at removing barriers to industrial development by lessening the burden of remediation costs and by permitting developers, lenders and investors to calculate those costs with more accuracy.

These new regulatory frameworks and programs integrate the following concepts in an attempt to encourage revival and development of urban industrial areas: reduction of clean-up standards using reality-based risk assessments; limitations of liability for purchasers and lenders; creation of express clean-up standards; establishment of a more efficient and responsive government review process; and provision of government incentives to attract developers, entrepreneurs, commerce and industry back to blighted urban areas. Innovative regulations and programs endeavor to address, in a coordinated and integrated manner, technical assistance from the regulators, liability assurances through covenants not to sue, and financial incentives, including tax abatements and credits.

The obstacles to the development of manufactured sites, and the emergence of innovative legislative programs and initiatives designed to overcome these barriers, are the subject of this chapter.

The nature of environmental liabilities

Overview

Environmental liabilities have their genesis in the social activism of the 1960s, which focused attention on protecting the earth's air and water. In the United States of America, the first environmental laws of note were enacted by the United States Congress. They regulated emissions of pollutants into the air, under the federal Clean Air Act (CAA),[1] and discharges of pollutants into waters of the United States, under The Federal Water Pollution Control Act (FWPCA).[2] Following the enactment of these first generation laws, Congress enacted the Resource Conservation and Recovery Act (RCRA),[3] which regulates hazardous waste and solid waste generation, handling, transportation and disposal for the "life" of this material, or from "cradle to grave."[4] RCRA broadly defines the terms: "solid waste"[5] and "hazardous waste."[6]

Shortly thereafter, perhaps the most well-known and far-reaching environmental statute was enacted by the United States Congress: the federal Comprehensive Environmental Response Compensation and Liability Act (CERCLA).[7] This law (and its state analogs) impose sweeping liability upon owners of property, operators and tenants of commercial businesses, transporters and disposers of "hazardous substances," as well as upon those who are causally responsible for the pollution. This statute, and the regulations promulgated by the United States Environmental Protection Agency (EPA) pursuant to it, defines the term "hazardous substances" broadly.[8]

This law is commonly referred to as "Superfund," because the statute provides for the collection of funds through taxation of the regulated community (industry and commercial private parties that handle oil and hazardous substances) in order to pay for clean-ups performed by the government. Such government clean-ups are necessitated when responsible parties fail to remediate the contamination.

The intent behind, and the most prominent features of, the United States environmental laws is consistent with that behind similar environmental laws in other Western industrialized countries. European environmental regulation is achieved through a combination of European Community (EC) Directives and domestic regulation. In addition to supplementing EC Directives, domestic regulations are often used as a means to implement them.[9] For example, Directive 91/156/EEC subjects all member states with waste disposal facilities handling hazardous wastes to a permitting requirement.[10] In order to comply with this directive, Italy implemented a regulation requiring, among other things, that facilities intending to handle a hazardous waste report "the possible risks associated with the substance; a description of the preventive measures to be taken by plant management; [and] the name of the individual at the facility responsible for safety . . ."[11] Likewise, Germany relies on its Federal Chemical Act to comply with EC Directive 67/548/EEC, which directs member states to "adopt ordinances and regulations pertaining to the classification, packaging and labeling of dangerous substances."[12] Germany has also recently adopted a broad solid waste law[13] that incorporates the American principle that the "polluter pays."

The United Kingdom attempted to deal with its environmental issue through the Control of Pollution Act of 1974 (CPA) – a massive piece of legislation covering waste disposal, water pollution, noise pollution, and various other environmental matters.[14] The CPA was insufficient to handle all environmental complexities. Therefore, it has been largely supplanted by the Environmental Protection Act of 1990 and the Water Resources Act of 1991. Similarly, following the privatization of their water industry, England and Wales now largely rely on the National Rivers Authority, which is authorized to recover clean-up costs from responsible parties, and to police its waterways.[15] Finally, instead of relying on medium-specific (e.g., air, water, soil, etc. . . . statutes,) France relies on a series of environmental statutes aimed at specific industries, including agriculture, chemical, and biotechnology.[16]

In the United States, the Superfund law is applied retroactively, which means that compliance and remediation responsibilities, as well as the resulting financial burdens, are imposed on private parties who presently own or operate a polluted site, transport waste to the site, generate wastes found at the site, and/or create or contribute to the pollution. The law is also applied to those who *formerly* had such status with reference to a contaminated site. These parties must address the pollution, without regard to the degree of their involvement in its handling, generation or disposal, and without regard to whether they deliberately or passively caused the problem. This central premise has been justified by one simple unassailable principle – "the polluters pay." Thus, while statutes were enacted decades ago in the United States with noble intentions, the statutory framework subsequently has been amended and expanded in order to impose broad liability on a wide class of parties, for a large range of damages. In addition, many states have emulated the federal government by enacting similar environmental statutes which govern pollution problems.

The existence of all of these statutes has greatly affected the ability to refurbish and develop former industrial and commercial polluted sites in urban areas. To complicate matters, common law doctrines concerning bodily injury and property damage (such as negligence, nuisance and strict liability for ultra-hazardous activities) have been interpreted by courts as also imposing liability on private parties for damages resulting from site contamination. This broad imposition of legal responsibility increases the financial and legal risks already inherent in urban redevelopment.

Comprehensive Environmental Response Compensation and Liability Act (CERCLA)

In the United States, the Superfund Statute[17] has the greatest impact on the revitalization and economic development of urban, industrial and commercial areas. This law was enacted during 1980, in the wake of the now infamous Love Canal incident. Superfund, and the many state laws modeled after it, serves two purposes.

The first and foremost purpose is the imposition of liability on the following widely inclusive classes of parties, commonly known as "Potentially Responsible Parties":

- Current owners and operators of a contaminated site or facility.
- Past owners and operators of the contaminated site or facility, who owned or operated at the time of disposal of hazardous substances.
- Transporters of hazardous substances to the site.
- Generators of the hazardous substances.

The liability is retroactive and is "strict, joint and several," meaning that a party's fault or lack of fault is not a consideration, and that each party is individually liable for the *entire* cost of the clean-up. Liability under this law extends to clean-up costs, as well as to damage to natural resources. The financial exposure to a Responsible Party can be staggering. Clean-up costs routinely exceed a quarter of a million dollars for a simple clean-up of a discrete area, and may easily exceed a million dollars for a complicated and extensive clean-up of a severe problem. Moreover, exposure on account of damages to natural resources (such as rivers, forests, lakes and groundwater) is essentially limitless. The measure of such damages may include the theoretical costs of restoration or replacement of a contaminated body of water or injured species of wildlife or plant life.

Superfund's secondary purpose is to authorize the government to collect a pool of money, or "Superfund," through taxation of the regulated industrial community. The government can draw on these funds to conduct clean-ups in situations where no responsible parties can be found or where liable parties fail to remediate the contamination.[18]

This same regulatory structure is often duplicated at the state level, where many states have enacted a "mini-CERCLA" statute modeled after the federal CERCLA statute. Specifically, forty-two states have clean-up statutes imposing retroactive liability, with forty-one of those states adopting the federal approach of strict liability.[19] States also generally have state level "superfunds," which are earmarked funds for clean-up related activities. In 1997, for example, forty-five states reported adding an aggregate $538.3m to their state clean-up funds.[20] The role that these state-level programs play cannot be overstated, given the fact that, from 1989 to 1997, states have reported completing more than 40,000 clean-ups.[21] The 1997 data also shows that states reported that they had more than 13,700 additional clean-ups underway.

In short, Superfund and its state analogs dictate which class of parties will bear the significant costs of clean-up. When considering the method of allocation of these costs among Potentially Responsible Parties, current and past property owners, facility owners, operators, tenants, and financial institutions must anticipate their roles and engage in a meaningful dialogue about the development of manufactured sites.

Resource Conservation and Recovery Act (RCRA)

In the early 1970s, there was a growing concern in the United States associated with the rising tide of waste materials being generated by an increasingly urban population. The

Resource Conservation and Recovery Act (RCRA)[22] was Congress's answer. This statute applies to facilities, which treat, store, or dispose of any hazardous waste or transporters of hazardous waste.

RCRA's potential impact on manufactured sites stems primarily from two sources. First, amendments passed in 1984 brought virtually all underground storage tanks (USTs) within RCRAs purview.[23] In fact, EPA estimates that there may be as many as 1.4 million tanks "containing petroleum products, hazardous wastes, or hazardous industrial chemicals" which pose a substantial threat of leaking.[24] The significance of RCRA for purchasers of sites containing USTs is that liability generally extends only to the current owner or operator of the tank.[25] Prospective purchasers should also be wary of the fact that it is often difficult for anyone cleaning up an underground storage tank leak to recover clean-up costs from former culpable owners.[26] In addition, tanks actively leaking hazardous substances can trigger CERCLA, drastically compounding the liability and clean-up issues.

The second area where RCRA's impact can be felt involves situations in which a site was once used as a manufacturing facility, which utilized hazardous wastes, or where a site was used as a landfill. For such a site, RCRA imposes closure requirements that must be satisfied before any redevelopment is allowed. RCRA regulations contain requirements which set standards for closing and cleaning facilities, sealing landfills, monitoring the groundwater, and even for monitoring landfill gas emissions. Any such redevelopment program in the United States may be further affected by the fact that each state is free to impose stricter standards than those set forth by the federal government. Accordingly, closure requirements, and hence redevelopment costs, can vary substantially from state to state.

Toxic Substances Control Act (TSCA)

The Toxic Substances Control Act of 1976 (TSCA)[27] empowers federal agencies to monitor and regulate various toxic substances that pose health and environmental hazards. Within the context of manufactured sites, lead paint and asbestos are two particularly relevant substances that are regulated under TSCA.

Lead paint hazards
Owners of residential buildings and child-occupied public facilities, such as schools and day-care centers, are required to "de-lead" their buildings by removing all chipping, flaking, or otherwise deteriorating lead paint.

"De-leading" of these facilities must be done by certified individuals[28] and may result in substantial renovation costs for private developers. Consequently, non-residential buildings that would otherwise be suitable for conversion to apartments or condominiums often remain non-residential, thus exacerbating the need for housing in economically deprived areas.

At this time, commercial properties in the United States that are neither residential nor primarily child-occupied are not required to be de-leaded. Owners of such properties are required to monitor the presence of lead paint, however, and must take remedial action upon a determination that lead chips or dust are emanating from the lead paint. Under this standard, buildings in which lead paint is present must be continually monitored for evidence that the lead paint is peeling, chipping, flaking, or otherwise deteriorating to such an extent that lead chips and/or dust may be released. Consequently, building owners and managers should be most concerned with the initial identification, and subsequent monitoring, of lead paint.

Asbestos hazards
Asbestos fibers pose a significant health risk if released into the environment. More specifically, individuals who inhale asbestos fibers over a prolonged period of time are at a

significantly increased risk of developing both lung cancer and a potentially fatal lung disease called asbestosis. Asbestos-containing materials (ACMs) in a building become dangerous when they are "friable," which means the ACMs are likely to release invisible asbestos fibers that are inhaled by the building's occupants. Environmental regulations issued under TSCA have established an "asbestos-in-place" program that allows building owners to simply monitor ACMs that do not pose a current threat of releasing asbestos fibers, rather than requiring building owners and developers automatically to remove non-friable (and, therefore, non-dangerous) ACMs.

Under the "asbestos-in-place" program, a properly accredited professional must identify and record the presence of all ACMs. After ACMs are found in a building, the building owner must institute policies and procedures that are designed to minimize contact with the ACMs, thus preventing – or at least minimizing – the risk that the ACMs will subsequently release asbestos fibers. Building owners are also required to train all workers, such as maintenance and service personnel, who can reasonably be expected to have an inordinate amount of contact with ACMs; more specifically, these individuals must be advised of the dangers posed by asbestos fibers, and they must be instructed on how to handle ACMs so as to prevent or minimize the risk of releasing asbestos fibers. Finally, building owners and managers should also (1) advise all building occupants of the presence of ACMs, (2) continuously and consistently monitor all ACMs within the building, and (3) insure that records regarding ACMs and other building activities, such as renovations that may damage or disrupt ACMs, are properly recorded.

Under most circumstances, building owners probably will not incur substantial expense simply because ACMs are present in a facility. Indeed, the most expensive task may be hiring properly accredited individuals who can advise the building owner or manager of the presence and, if applicable, condition of all ACMs within a building. The costs associated with this task, as well as the cost of training at-risk employees for asbestos-related problems, should not be so expensive as to preclude the continued use of an existing industrial or commercial building.

On the other hand, renovating or otherwise modifying a building for a new or additional use may require the removal of all ACMs in order to avoid the risk of inhaled asbestos fibers, if such renovation will lead to widespread disruption of ACMs. Therefore, the identification of ACMs which will be released into the ambient air of the facility during renovation must be considered when budgeting for the redevelopment of a manufactured site. Furthermore, the costs associated with removing the ACMs from the building in a safe manner and replacing the ACMs with an approved insulator or fire retardant can be significant. Nevertheless, developers who wish to renovate a building that contains ACMs may have no option other than this potentially expensive ACM removal and replacement, because the "asbestos-in-place" program is available only if the ACMs are not subject to any activity that may lead to the release of asbestos fibers. Of course, developers faced with this situation may elect to compartmentalize strictly those portions of a building to be renovated, thus potentially reducing the amount of ACMs that must be removed. Before such a course of action is undertaken, however, developers are well advised to ask the appropriate environmental authorities whether such limited renovation will correspondingly limit the scope of required ACM removal.

Clean Air Act (CAA)

By imposing air emission restrictions, the Clean Air Act (CAA)[29] can hinder the formation of manufactured sites when the redevelopment project involves new or expanding industrial enterprises in the United States. If projected emissions are expected to exceed the allowable local or national limits for specified air pollutants, then the project will be forced to relocate.

Air pollution control begins under the CAA by the imposition of national limits on the allowable concentration of certain air pollutants.[30] In areas that have not yet exceeded these limits, the CAA further imposes a limit on the allowable incremental pollution increase created by a project. Such areas are deemed "prevention of significant deterioration" (PSD) areas.[31] Thus, a potential manufactured site may be in violation of the CAA because it leads to the significant deterioration of the ambient air, even if it does not violate the national air standard. For example, Disney American sought to build a theme park in northern Virginia. The project was to consist of a 3,000-acre theme park, which was to be located some 25 miles from Washington, DC.[32] The project was nearly approved by the legislature of the State of Virginia when the "[local governmental bodies] . . . were forced to reconsider their approval once they realized that the resulting traffic congestion and air pollution would have used up most of the region's available incremental increase in emissions allowable under the CAA."[33] Potential developers should note, however, that PSD provisions of the CAA may actually have the effect of limiting urban sprawl and encouraging site remediation. This is due to the fact that PSD incremental allowances are greater for more urbanized areas, thus affording industry more lenient emission standards when choosing to redevelop a manufactured site in these urbanized areas.

Disincentives to redevelopment

Environmental liabilities are unintended disincentives to redevelopment of contaminated urban sites. They present significant obstacles to the key parties to redevelopment: current owners, potential purchasers and lending institutions. A lack of teamwork and coordination among developers, lenders, government agencies, community groups, environmental engineers, landscape architects and attorneys often exacerbates the problem.

Current owners/potential purchasers

Current owners and potential purchasers of an industrial or commercial site that is polluted are alarmed by the existing and/or potential liabilities associated with on-site pollution and/or migration of pollutants from abutting property. They equate these liabilities with unwanted expenditures, such as paying for a mandatory clean-up under Superfund or buying required pollution control equipment under the CAA or WPCA. Moreover, they view such expenditures as potentially significant, difficult to budget for, and an unnecessary and extraordinary drain upon operating expenses.

Their concerns, unfortunately, are well grounded. Regardless of fault, owners and other Potentially Responsible Parties must clean up according to strict standards and requirements set by the laws governing contaminated urban sites. The remediation of a manufactured site is dictated or governed by detailed environmental regulations, which require in-depth engineering, chemical, toxicological, and statistical studies, and a variety of permits prior to implementing a clean-up action. Governmental review is typically required. Permits and approvals by administrative or environmental agencies often must be obtained prior to proceeding to the next stage of work. Delay by government agencies can have the effect of postponing the scheduling and completion of such a project, which means that the environmental work necessary to create or develop a manufactured site can take years to complete. In sum, the high potential for environmental expenses at a manufactured site should always be considered when devising and revising a renovation or development budget.

Lending institutions

Environmental liabilities also create problems for lending institutions, which otherwise might provide financing for manufactured sites. These liabilities concern financing institutions because they call into question the value of using contaminated property as collateral for any necessary loan. Banks and other lenders reason that environmental liabilities may require the incurrence of clean-up costs by the borrower, reducing net operating income of the subject property, reducing the property's fair market value, and rendering the property virtually worthless as collateral for a loan. Thus, lending institutions, when considering whether to finance redevelopment, view environmental liabilities as a threat to the subject property's value as collateral. In addition, lending institutions are concerned about the effect of potential environmental liabilities upon the borrower's ability to repay the loan. Thus, environmental liabilities can dissuade a usually necessary party, the lender, from participating in the transformation of an unproductive contaminated urban site.

Even when contaminated property retains sufficient fair market value to cover a loan, environmental liabilities remain of major concern to financing institutions, because they believe that environmental liabilities may create an obstacle to any necessary foreclosure proceedings. For example, environmental statutes often include "lien" provisions, allowing the relevant government party to place a lien for multiple damages on contaminated property in order to secure recovery of its clean-up expenditures. Such a lien provision may even give the government a priority security interest over the lending institution (i.e., a "super lien"). Many lenders wish to avoid a subordinate creditor position or lien complications, which may create an obstacle to a foreclosure proceeding. As a result, they may decide not to provide any financing whatsoever for a contaminated site. Further, lending institutions fear that, upon foreclosure, they may incur liabilities as an "owner" or "operator" of the property under Superfund. All of these risks negatively influence the lending community, which already questions how a borrower, saddled with extraordinary environmental costs, will be able to meet the loan repayment obligation. In sum, environmental liabilities affect the collateral for a loan, the borrower's ability to comply with periodic payment schedules, the availability of foreclosure proceedings, and the expected return on the loan. Consequently, lending to a developer of a manufactured site that is contaminated is a far less attractive proposition than lending to a purchaser of a pristine site. To address these problems, statutory and private sector incentives have recently been implemented in the United States in order to stimulate development of these sites.

Community members

Although certain community groups voice an interest in promoting the clean-up and redevelopment of neighborhood commercial and industrial areas, they understandably expect some assurance that such development of manufactured sites will adequately protect their health and the environment. In addition, some communities prefer the replacement of industrial manufacturing facilities with commercial facilities, residential buildings or neighborhood parks and recreational spaces, which they perceive as less dangerous to the public health and as less harmful to the environment. Unfortunately, this often creates conflicts between potential developers and community groups who want the government to ensure the environmental safety of their neighborhoods. Many federal and state statutes contain provisions which allow for public comment and participation in the siting, use, design and renovation of manufactured sites.[34] Developers of manufactured sites must work cooperatively with their attorneys, environmental engineers, landscape designers and such community groups in order to respond to these concerns.

Initiatives to minimize liability and accelerate manufactured site redevelopment

During the late 1990s, government and industry focused on limiting the sweeping nature of environmental liabilities in order to encourage the development of manufactured sites.

Representatives of a wide variety of interests took on the challenge of attempting to make clean-up costs more manageable by offering incentives and undertaking certain initiatives. The continuing evolution of these initiatives offer all concerned parties the ability to meaningfully participate in the dialogue and process concerning the development of manufactured sites.

Superfund

Superfund liability discourages property transfers because liability is imposed, without regard to fault, among "current owners" of sites.

Brownfields

Recently, government programs have begun to address this problem by enacting laws which exclude from liability new purchasers or new owners of contaminated sites, if the new owners or purchasers did not cause the contamination and did not own or operate the site during the time-frame when the contamination was released. Of course, this protection is only available if the site is cleaned up. The United States Congress has drafted proposed legislation addressing this policy of encouraging the re-development of polluted sites in urban areas. This initiative is referred to as "Brownfields" legislation, because it encourages the productive use of manufactured sites, rather than new development of open spaces – "Greenfields."[35]

EPA's Brownfields agenda is actively removing many barriers to redevelopment. In addition to providing pilot site grants to seventy sites nationwide, the agency has focused on creating partnerships with state and local government, as well as with other federal agencies.[36] EPA has also encouraged certain redevelopment projects, known as "pilot sites," in order to craft better and broader policies and strategies; for example, in Louisville, Kentucky, lessons learned during clean-up and redevelopment of the Ni-Chro Plating site helped establish policies for Louisville's city-wide Brownfields program. Similarly, construction of the federal courthouse in Sacramento, California, has served as a blueprint for policies affecting the remaining 240 areas of Southern Pacific's rail yard.[37]

Many states, such as California, Massachusetts, New York, and Washington, have already enacted beneficial Brownfields legislation. For example, the Commonwealth of Massachusetts recently enacted a law, which modified its Superfund liability scheme.[38] Pursuant to these changes, a purchaser or developer, under certain circumstances, can acquire an ownership interest in an already contaminated site, and later be protected from liability when a permanent solution to the contamination is achieved. The new owner or tenant is exempted both from liability under the state's Superfund statute, and from liability under common law doctrines, for clean-up costs and property damage, to both the government and private third-parties. This achievement is of monumental importance for the redevelopment of contaminated industrial and commercial sites, because potential purchasers of such sites are encouraged by the statute to become current owners or tenants. Such legislative initiatives are promising, albeit controversial and still relatively unusual.

Memoranda of Agreement

One of the most lamented liability features of federal and state Superfund law is the fact that remediation performed under the auspices of a government agency does not absolve a would-be purchaser from liability. Further, clean-up agreements between potential purchasers and the federal government are not binding on state environmental agencies. In an effort to ease purchasers' fears of dual accountability, the EPA encourages coordination between federal and state agencies via agreements known as Memoranda of Agreement (MOAs). In August 1997, the EPA issued the final draft MOA guidance detailing the guidelines that will be used to align federal and state statutory enforcement efforts. This guid-ance provides that the "EPA will not exercise cost recovery authority and does not gener-ally anticipate taking CERCLA removal or remedial action at sites covered by a MOA except under limited circumstances detailed in the guidance."[39]

Lender liability
The United States federal government and various state governments are currently taking more creative and aggressive actions to encourage lenders to participate in the develop-ment of manufactured sites. Environmental liabilities are being curtailed and new incen-tives are being developed[40] to allow financing institutions to make loans on contaminated property without necessarily incurring liability as an "owner/operator" pursuant to fore-closure proceedings on the contaminated property. One of the most significant steps Con-gress has taken to ameliorate the counterproductive liability effects of federal environmental laws is the enactment of EPA's policy of protection for lenders who were not actually involved in the management of a contaminated site.[41] The public policy stressed in the passage of such legislation encourages lenders to make funding decisions based on the merits of the investment itself, rather than on liability considerations in the event of a foreclosure on contaminated property. For the would-be developer, this innova-tion should facilitate the ability to obtain development funds, as well as foster the creation of financing alternatives. These favorable developments in the law will be critical determi-nants for whether a contaminated site is redeveloped.

Many states also now realize the benefits of shielding lenders from liability. For example, the state of Massachusetts recently passed a law which virtually assures that non-culpable lenders will be exempt from environmental liabilities.[42]

Potential purchaser agreements
EPA has also sought to encourage contaminated site redevelopment by entering into clean-up agreements with prospective developers. These Prospective Purchaser Agree-ments (PPAs) generally call for a redeveloper to clean up the contaminated site to a prede-termined level, in exchange for a promise from the EPA not to sue for further clean-up costs that may be required in the future. The value of such an agreement, to a purchaser of contaminated sites, is obvious. By removing some of the uncertainty surrounding remediation, purchasers and lenders are much better able to evaluate the viability of the underlying investment.

Although EPA's PPA guidelines have existed since 1989, they have had limited success. Much of the failure of the earlier PPA initiative stemmed from the fact that the guidance was ambiguous and required that the EPA receive a "substantial benefit" from the prospective purchaser.[43] This requirement often included a clean-up procedure, as well as direct payments to the EPA.[44] However, in 1995 the EPA issued a new guidance for PPAs, which only required an "indirect public benefit in combination with a reduced direct benefit to the EPA."[45] The EPA believes this modification to the "benefit" requirement will drastically lessen the burden on purchasers seeking protection for unknown future clean-up costs.

Risk-based clean-up costs: "how clean is clean?"
Clean-up costs tend to rise exponentially as the standard of required remediation becomes more stringent. Therefore, if an acceptable level of human safety can be determined, then substantial financial benefits can be realized by limiting the amount of clean-up to only that which is required to achieve that predetermined safety level. The EPA and many states have recently embraced this idea by recognizing the value of risk-based clean-ups.[46]

In the past, risk-based clean-ups were based on the idea that the level of clean-up should be established so that the amount of risk to humans remains constant, regardless of the future intended use of the site. Under this standard, the scope and degree of reme-diation was, at least implicitly, tied to the specific site's level of human activity. For example, human exposure to contaminants is, generally, greater at a residential site than at a site slated for commercial or industrial use. Thus, to maintain an across-the-board low risk level, a residential site requires more contaminant removal than an industrial site. Fur-thermore, to ensure that this predetermined risk level is never exceeded in the future, the deed to the underlying property can be amended to include the same use restrictions that were utilized to determine the risk when the risk assessment was performed.

Although CERCLA has always permitted the EPA to consider *possible* future land use when setting the required level of clean-up, the EPA has traditionally assumed the future use of land to be residential.[47] Accordingly, the level of clean-up generally required was based on the "worst-case" residential land use. The EPA, in making this "worst case" sce-nario determination, considers a hypothetical 16-kilogram child who ingests 200 milli-grams of soil every day for five years.[48] The financial ramifications of this unrealistic assumption can be significant to a developer of a manufactured site slated to be used for only industrial purposes.

In an effort to bring clean-up assumptions in line with reality, the EPA, in 1995, restated its commitment to the idea that "actual" future land use should be taken into account when choosing a remedy.[49] EPA further proposes to use deed restrictions and conventional physical barriers[50] to ensure that the maximum risk level is never exceeded. Prospective developers of manufactured sites should note, however, that CERCLA requires *all* clean-ups to comply with the standards of *all* applicable state and federal laws.[51] Therefore, although the EPA's use of risk-based standards tailored to the anticipated future use of the land may eliminate unnecessarily stringent federal requirements, concerns arising from state-level clean-up standards in reliance upon future land use will persist. Further, competing interests in adjusting standards to more closely match the expected use of a particular manufactured site could raise concerns, in some instances, about the adequacy of protection for human health and the environment.

Privatization

Currently, environmental clean-up regulations are undergoing amendment in an effort to simplify the clean-up process, and reduce the time and money needed to achieve a clean-up. These regulations typically first require comprehensive engineering studies, which are evaluated and approved by the government.

Some progressive regulations are based on "privatizing" the clean-up process. The idea is to shift oversight of, and ultimate responsibility for, the clean-ups from the government to professional environmental engineers, who are specially licensed by the state in which they practice, and who are experts in evaluating and dealing with hazardous substances. These licensed environmental engineering professionals are required, by law, to be hired by Potentially Responsible Parties dealing with the clean-up and development of a manu-factured site. The licensed professionals also have the responsibility of filing their evalu-ations, proposed designs, plans and progress reports with the government. This process should incorporate the legal requirements which must be anticipated by landscape archi-tects, engineers, developers, and architects, as they design and plan for the use of site areas where soil excavation is prohibited or limited (due to compliance with regulations), or where impact to groundwater or air emissions must be considered. The purpose of this procedure (which relies upon the work and professional opinions of these private consul-tants) is to eliminate the necessity of obtaining governmental approvals, including the determination by the government that no further action is required. Privatization increases the need for coordination of all professionals involved at a manufactured site, including the attorneys, but it is expected to result in more speedy clean-ups and more accurate estimates for the completion of any necessary remediation. However, this initiative avoids the threshold environmental liability issue: prospective purchasers often do not want to incur any potential significant liability – whether to the government or a private party – for clean-up costs.

Privileged and confidential audits

Environmental audits[52] are often the only vehicle by which companies are able to discover and remedy environmental violations. However, the problem lies in the fact that information contained in the audit reports can be used by the EPA in a suit for non-compliance against the very company that voluntarily performed the audit. Purchasers and redevelopers are also at risk if they order an audit that indicates the site violates some environmental regulation. The public policy consideration against allowing violations discovered in an audit to be completely privileged ensures that companies cannot avoid environmental responsibility altogether by hiding behind a privileged audit shield. Hence, the challenge that the government faces is that of encouraging companies to monitor themselves and correct environmental violations, while not providing industry wholesale immunity for illegal acts.

To meet this challenge, the EPA adopted, in December 1995, a policy aimed at encouraging companies to perform voluntary environmental audits by drastically reducing the penalties for violations discovered via such audits.[53] Most states are now following suit by either passing or considering laws that create an "environmental self-audit privilege."[54]

Private party indemnifications

For many years, sophisticated commercial purchasers of real estate have obtained, from sellers, indemnities for environmental liabilities. Additionally, they routinely require an environmental investigation before finalizing the purchase of any commercial or industrial property. These purchasers obtain such reports in order to identify environmental problems, and utilize the information in order to negotiate indemnity provisions with the sellers. A typical comprehensive environmental indemnity provision is predicated on two guiding principles:

- the seller agrees to pay for all clean-up costs, judgments, fines, and other damages incurred by the buyer on account of the existing contamination; and,
- the seller agrees to defend the buyer, or pay for the buyer's defense costs, in connection with any response to environmental claims brought against the buyer by the government or private third parties.

Comprehensive seller indemnity provisions encourage purchasers to move forward on a real estate transaction, even when the subject property is extremely contaminated. However, this strategy has its shortcomings. The indemnity provision, without security, is only as good as the indemnifying party's ability to meet the financial obligations set forth in the indemnity agreement. For this reason, such indemnity agreements also often require collateral and/or the establishment of escrow funds in favor of the purchaser. Such funds must be used for property clean-up and/or to defray potential liability costs associated with the pollution.

Moreover, private party indemnity provisions only affect the liabilities and rights as between those parties. Such provisions do not eradicate a purchaser's or seller's liability to the government or to other private third parties (such as abutting property owners). Therefore, a purchaser could obtain a comprehensive indemnity from the seller, then acquire the property, and still be sued by the government for clean-up and/or be sued by private third parties for diminution of the value of their abutting properties on account of the migration of the pollution or the stigma to their properties. Of course, private third parties may also make claims against the current owner for bodily injuries suffered as a result of exposure to the contamination. This scenario creates an untenable position if the seller has no financial resources to make good on its indemnity to the buyer, and inadequate collateral or escrow funds have been maintained. Therefore, indemnity agreements between sellers and purchasers, while useful, are not a foolproof solution for manufactured sites.

Remediation cap and third-party insurance

Recently, the insurance industry has responded to the open-ended risks to developers and owners which are presented by contamination. A wide variety of insurance packages and options are now available to property owners who face environmental liabilities associated with a manufactured site. These insurance options are relatively new, because insurers have traditionally been very reluctant to provide coverage for environmental liabilities. The more prominent examples of environmental insurance policies include "clean-up cost cap" coverage, "pollution legal liability select" coverage, and "storage tank pollution" insurance.

"Clean-up cost cap" coverage protects property owners and operators from financial losses that might occur when environmental clean-up costs exceed expected remediation expenditures. This type of coverage insures against cost overruns, so property owners and operators who know they will be required to participate in an environmental remediation project should explore clean-up cost cap coverage as a way of minimizing the risk and uncertainty associated with the financial impact of clean-up operations. Generally, the coverage begins after the planned clean-up expenditure, plus a pre-determined "buffer," is exceeded. For example, a property owner or operator who expects to spend $1 million in environmental clean-up might purchase a clean-up cost cap policy with a $100,000 "buffer." The clean-up cost cap coverage will begin after the property owner or operator spends $1.1 million, which represents the $1 million of expected expenditures plus the $100,000 "buffer."

"Pollution legal liability select" insurance provides a property owner or operator with a menu of options from which to create an environmental insurance policy that is tailored to the risks present at a particular site. These coverage options include protection against third-party claims for off-site bodily injury or property damage, as well as claims to recover environmental remediation costs. Property owners and operators may also purchase coverage for on-site clean-up activities.

Owners and operators of manufactured sites with underground storage tanks (USTs) should consider "Storage Tank Pollution Insurance," which provides coverage for bodily injury and property damage to third parties resulting from tank leaks. These policies can also provide on- and off-site clean-up coverage, and they can be utilized to meet state and federal regulatory requirements regarding the existence and use of USTs.

Finally, building owners who are faced with lead and/or asbestos abatement requirements may purchase "lead abatement liability insurance" and/or "asbestos abatement liability insurance" which, respectively, insure against property damage and third-party bodily injury incurred during lead and asbestos abatement projects. Similarly, "abatement umbrella" coverage provides additional insurance for liabilities arising from lead and/or asbestos abatement projects.

In sum, developers, facility owners and managers should be aware that there are a variety of insurance options available, and should consult with an insurance broker in order to evaluate their specific insurance needs.

Tax credits and tax deductions

The income tax treatment of environmental remediation and related expenditures is a key element in determining the potential cost and budget of any proposed environmental action or remedial alternative. In weighing its remedial options, a property owner must determine whether any such remediation expenses are currently deductible, capitalized and deducted over the life of the asset, or not deductible at all. In some states, the property owner's options now include a credit against tax for certain properly incurred expenses.

Environmental expenditures which are treated as ordinary and necessary business expenses may be used to offset current income subject to tax.[55] Capital expenditures, on the other hand, must be recovered through depreciation deductions over the useful lifetime of the asset.[56] This means that the full cost of the expenditure cannot be deducted from income in the year that it is made, but must be spread out over time. A credit, in contrast, serves to reduce, dollar for dollar, the tax liability of a taxpayer, and, consequently, is the most prized tax benefit of all.

Although the tax law historically has been used to encourage desired behavior and discourage undesirable behavior, the federal tax policy was for many years at odds with federal environmental policy, which encouraged privately funded remedial activities. Instead of allowing a deduction for the cost of environmental clean-ups, the United States Internal Revenue Service (IRS) generally treated such costs as permanent improvements to property that had to be capitalized and deducted over time.[57] The IRS supported its conclusion by arguing that expenditures to clean up property increased the value of the property, and thus were capital in nature. In every case, the IRS easily concluded that the value of the taxpayer's property had increased because it compared the value of the taxpayer's property after it had been cleaned up with the value of the property as contaminated.

After years of comments and complaints from taxpayers and members of Congress, the IRS changed its position in 1994, specifically ruling that costs: (1) of evaluating soil and groundwater contamination problems that occurred during the taxpayer's operation; (2) of remediating the soil; and (3) of ongoing groundwater treatment expenses were deductible as ordinary and necessary business expenses.[58] After this ruling, the new focus of the IRS inquiry is now upon the life, use, and value of the property *before* the contamination and after the clean-up. If the environmental expenditure restores contaminated property to its previous condition, the action would be considered a "repair," which is a deductible business expense. However, when the value of the property is materially increased, as compared with the status of the asset prior to the condition necessitating the expenditure, the expenditure will generally be considered an "improvement" that must be capitalized.

Unfortunately, this ruling only addressed clean-up costs incurred by the same taxpayer who contaminated the property in the first place, rather than by someone who acquired a previously contaminated property. This ruling provided no relief to the potential purchaser of contaminated property, who would have to capitalize any expenditures to clean up property that was contaminated by a previous owner.

To provide an incentive to the private sector to clean up these sites, particularly those located in economically distressed communities, Congress recently enacted a law[59] that permits a current deduction for certain remediation costs, paid or incurred in connection with the abatement or control of environmental contaminants with respect to qualified sites. To qualify, the site must satisfy land use, geographic and contamination requirements. The use requirement is met if the property is used in the taxpayer's trade or business or for the production of income. The geographic requirement is satisfied if the property is located in one of the following areas: (1) EPA "Brownfields" Pilot areas designated prior to February 1997; (2) census tracts where 20 percent or more of the population is below the poverty level; or (3) any Empowerment Zone or Enterprise Community (and any supplemental zone designated on December 21, 1994). To satisfy the contamination requirement, hazardous substances must be present or potentially present on the property. The criteria for eligibility permits both rural and urban sites to qualify for this tax incentive.

This legislation corrects, at least in certain cases, the contradictory tax treatment of clean-up costs for buyers and sellers of contaminated property. If a person contaminates property and then cleans it up before a sale, the costs of that clean-up are expendable. However, prior to this legislation, if a purchaser cleaned up a contaminated site, those costs had to be capitalized under the assumption that the purchase price was reduced to reflect the costs of clean-up. This federal law expires after three years, meaning that only eligible expenses incurred or paid from the date of enactment until January 1, 2001 qualify for deduction. The trend in allowing more deductions for a wider range of environmentally related expenditures continued when the IRS recently allowed for the deduction of consulting and legal fees associated with the proposed remediation of a site.[60] This IRS decision, which reversed an earlier ruling, held that the consulting and legal expenses did not add value or create an asset producing long-term benefits. Therefore, these expenses should be deductible as ordinary and necessary business expenditures. Based on long-standing tax regulations, the IRS has drawn the line, currently, at the expenditure for investment in remedial equipment or facilities.[61] According to the IRS, such equipment or facilities (e.g., a wastewater treatment facility, monitoring wells, air-stripping devices) add to the value or substantially prolong the useful life of property owned by the taxpayer.

At the state level, many innovative programs have been enacted to motivate private environmental clean-up measures. For example, in the state of Massachusetts, a recently enacted law[62] provides certain taxpayers with, among other measures, tax credits for environmental response costs incurred on contaminated business properties in economically distressed areas. If qualified, these taxpayers may take a tax credit of up to 50 percent of their costs incurred prior to January 1, 2005. Also, municipalities may establish agreements with eligible taxpayers for the abatement of real estate taxes on sites zoned for commercial or industrial uses.[63] The purpose of the agreement is to enable environmental response actions to continue on such sites and to encourage redevelopment in the community.

To encourage environmental remediation, the tax treatment of various clean-up options should correspond to the spectrum of clean-up outcomes, giving more favorable treatment to the choices that are more preferable from an environmental point of view. Accordingly, the costs of immediate remediation should be expensible, in order to provide an economic reason to prefer prompt clean-up over delayed clean-up. This incentive justifies the expensing of pollution control equipment that otherwise would have to be capitalized. Tax planning has long been recognized as extremely important in determining whether to perform clean-ups of hazardous waste sites. The tax treatment of necessary expenditures can dramatically alter the total cost of multi-million dollar clean-up operations and encourage certain behavior, be it environmentally friendly or otherwise. Accordingly, the trend should continue in the direction of providing tax incentives that will encourage private clean-up and avoidance of environmental contamination.

Conclusion

The development of manufactured sites requires a comprehensive understanding of the legal framework which governs land use and remediation. Prospective purchasers and developers, landscape architects and design professionals, and environmental engineers should include environmental lawyers on their team at the earliest opportunity. The redevelopment process necessitates cooperative alliances at every juncture. Land use planning, community activity, government oversight, and the compliance with the law concerning site remediation are factors that underscore the prudence of retaining a team of professionals to aid in the facilitation of the necessary collaboration efforts inherent in the successful development of manufactured sites.

Acknowledgement

A special note of appreciation is due to Jonathan Lindsay, Harvard Law School Class of 2000.

Notes

1 42 U.S.C. §§ 7401 *et seq.* (1994).
2 33 U.S.C. §§ 1251 *et seq.* (1994).
3 42 U.S.C. § 6901 *et seq.* (1994).
4 The term "cradle to grave" has been used by commentators to describe the all-encompassing nature of RCRA's regulation of solid and hazardous wastes.
5 RCRA defines the term "solid waste" to include: "any garbage, refuse, sludge from a waste treatment plant, water supply treatment plant, or air pollution control facility, and other discarded material, including solid liquid and semi-solid, or contained gaseous material resulting from industrial, commercial, mining, and agricultural operations, and from community activities . . ." 42 U.S.C. § 6903(27) (1994).
6 RCRA defines the term: "hazardous waste" to include: "a solid waste, or combination of solid wastes, which, because of its quantity, concentration, or physical, chemical, or infectious characteristics, may (A) cause or significantly contribute to an increase in mortality or an increase in serious irreversible, or incapacitating reversible illness; or (B) pose a substantial, present or potential hazard to human health or the environment when improperly treated, stored, transported, or disposed of, or *otherwise managed.*" 42 U.S.C. § 6903(5)(A)–(B) (1994) (emphasis added).
7 42 U.S.C. §§ 9601 *et seq.* (1994).
8 What makes CERCLA's definition of "hazardous substance" so broad is the fact that it incorporates by reference all "hazardous substances" designated as such under the FWPCA, CAA and RCRA. See 42 U.S.C. § 9601(14) (1994). With over 700 materials listed as "hazardous substances," the issue of identifying a "hazardous substance" is further complicated by the fact that any waste material which contains a "hazardous substance," regardless of the amount, is itself a "hazardous substance" for CERCLA purposes. See *United States* v. *Alcan Aluminum Corp.*, 964 F.2d 252, 259–61 (3d Cir. 1992).
9 See Trevor Adams College of Law, London, "Environmental Law in the European Communities," in *Comparative Environmental Law & Regulation* (Nicholas A. Robinson ed., Oceana Publications, Inc.) (October 1997).
10 Ibid. at 18.
11 Vincenzo Sinisi *et al.*, "Environmental Law of Italy," in *Comparative Environmental Law & Regulation*, 18 (Nicholas A. Robinson ed., Oceana Publications, Inc.) (November 1996).
12 Dr Herbert Posser *et al.*, "Environmental Law of Germany," in *Comparative Environmental Law & Regulation*, 24 (Nicholas A. Robinson ed., Oceana Publications, Inc.) (April 1997).
13 [Waste Avoidance, Recycling and Disposal Act] (effective October 7, 1996).
14 See Ian Doolittle, "Environmental Law of the United Kingdom," in *Comparative Environmental Law & Regulation*, 11–12 (Nicholas A. Robinson ed., Oceana Publications, Inc.) (November 1996).
15 Ibid. at 15.
16 See Andre Soulier *et al.*, "Environmental Law of France," in *Comparative Environmental Law & Regulation*, 25–8 (Nicholas A. Robinson ed., Oceana Publications, Inc.) (November 1996).
17 42 U.S.C. § 9601 *et seq.* (1994).
18 The taxing authority funding the Superfund trust expired in December 1995. Although several bills have been introduced to reinstate the Superfund taxes, as of yet none have made it through Congress. At current pace, the Superfund trust will be funded for only two more years. See Carol M. Browner, Administrator U.S. Environmental Protection Agency Statement Before the Subcommittee on Water Resources and Environment, U.S. House of Representatives (May 12, 1999), <http://www.epa.gov/superfund/new/congress/05-1299.htm> (visited October 6, 1999).

19 See Linda K. Breggin *et al.*, "State Superfund Programs: An Overview of the Environmental Law Institute's (ELI's) 1998 Research," 4 *Alb. L. Envtl. Outlook* 1, 7 (Winter 1999).

20 Ibid. at 2.

21 Ibid. at 4.

22 42 U.S.C. §§ 6901 *et seq.* (1994).

23 *See* 42 U.S.C. § 6991–6992 (West. 1994).

24 Charles Bartsch and Elizabeth Collaton, *Brownfields: Cleaning and Reusing Contaminated Properties* 14 (1997).

25 See 42 U.S.C. § 6991(b)(6)(A) (1994) (stating that the owner or operator shall be liable for costs incurred due to UST petroleum releases); see also § 6991(b)(C) ("No indemnification, hold harmless, or similar agreement or conveyance shall be effective to transfer from the owner or operator of any underground storage tank . . . to any other person the liability imposed under this subsection."). Although current owners will generally not be able to escape liability, they remain free to bring suit against former culpable owners.

26 The United States Supreme Court held that owners of a UST cannot recover clean-up expenses from former owners under RCRA's citizen suit provision. See *Meghrig* v. *KFC Western, Inc.*, 116 S.Ct. 1251 (1996). If, however, the tank presents an "imminent hazard," then liability will extend to anyone "who has contributed" to the situation. See 42 U.S.C. §§ 6972(a)(1)(B), 6973(a).

27 15 U.S.C. § 2601 *et seq.* (1994).

28 See 15 U.S.C. § 2682(a) (1994).

29 42 U.S.C. §§ 7401 *et seq.* (1994).

30 The EPA must establish nationally uniform ambient air quality standards (NAAQS) for pollutants that "may reasonably be anticipated to endanger public health or welfare." See 42 U.S.C. § 7047(d)(1)(A).

31 Facilities in such areas must show that their emissions will not violate any NAAQS or incremental allowance. See 42 U.S.C. § 7475(a)(3). Allowable incremental increases are based on the existing ambient air quality in the PSD area and the percentage of increase allowed over the baseline. See 42 U.S.C. § 7479(4).

32 Bartsch and Collation, *supra* n. 24, at 51–2.

33 Ibid. at 52.

34 CERCLA provides for public participation in the clean-up process consisting of: notice to potentially affected persons, an opportunity for public comment on a proposed clean-up plan, and a requirement that the EPA respond to all significant comments, criticisms, or new information presented by the public. See 42 U.S.C. § 9613(k)(2)(B) (1994). Likewise, forty-seven states reported having some type of public participation in their clean-up processes. See Breggin *et al.*, *supra* n. 19, at 3. Massachusetts, for example, requires that a public meeting be held upon receiving ten or more petitions for potentially affected residents. See Mass. Gen. Laws ch.21E, § 14 (1994). In addition, Massachusetts requires that the proposed clean-up plan be revised to reflect the comments it receives at this meeting (ibid).

35 USEPA defines "Brownfields" as "[a]bandoned, idled, or under-used industrial and commercial facilities where expansion or redevelopment is complicated by real or perceived environmental contamination." <http://www.epa.gov/swerosps/bf/glossary.htm#brow.htm> (last updated September 30, 1997) (visited March 22, 2000). "Greenfields" is a term that refers generally to undeveloped non-urban areas. See *The Brownfields Economic Redevelopment Initiative: Proposal Guidelines for Brownfields Assessment Demonstration Pilots*, <http://www.epa.gov/swerosps/bf/htm1-doc/appbook.htm> (visited October 8, 1999).

36 See generally, Brownfields National Partnership Action Agenda, <http://www.epa.gov/swerosps/bf/htm1-doc/97aabref.htm> (last updated March 18, 1998) (visited March 22, 2000)

37 Editts Pepper, *Lessons from the Field, Unlocking Economic Potential with an Environmental Key*, Northeast Midwest Institute (1997).

38 See generally, Brownfields Act: § 206 of the Acts of 1998 (as codified in Mass. Gen. Laws, ch.21E).

39 See Final Draft Guidance for Developing Superfund Memoranda of Agreement (MOA) Language Concerning State Voluntary Clean-up Programs (visited July 24, 1998) <http://www.epa.gov/swerosps/bf/html-doc/sfmoamem.htm>

40 See *infra* n. 41 and accompanying text; see also *infra* n. 42 and accompanying text.

41 In *U.S.* v. *Fleet Factors Corp.*, 901 F.2d 1550 (11th Cir. 1990), the court held that the lender was liable as the "operator" of the facility because "its involvement with the management of the facility is sufficiently broad to support the inference that it could affect hazardous waste disposal if it so chose." In 1992, the EPA adopted the lender liability rule that specifically defined the scope of activities which a lender could undertake without "participating in management." See 57 Fed.Reg. 18, 344 (1992). However, this rule was subsequently struck down in *Kelly* v. *EPA*, 15 F.2d 1088 (D.C. Cir. 1994). In 1995, EPA announced that it would follow the now-vacated lender liability rule as a matter of policy. Finally, this policy of not pursuing lenders who do not "participate in management" was codified in the Asset Conservation, Lender Liability, and Deposit Insurance Protection Act of 1996 (Pub.L. 104-208, 110 Stat. 3009-462, Div. A, Title II, Subtitle E (Sept. 30, 1996) (42 U.S.C. §§ 6991b, 6991b note, 9601, 9601 note, 9607).

42 *See* Mass. Gen. Law ch.21E, § 2 (as amended by the Brownfields Act: Section 206 of the Acts of 1998) (expanding and clarifying the liability exemptions for lenders who satisfy the duties imposed on them by the Act, and replacing the "participates in management" liability standard with a more culpability based causation standard).

43 See Superfund Program; De Minimis Landowner Settlements, Prospective Purchaser Settlements, 54 Fed.Reg. 34,225 (1989).

44 Todd S. Davis and Kevin D. Margolis, *Brownfields: A Comprehensive Guide to Redeveloping Contaminated Property* 42 (1997); See also Joel B. Eisen, "Brownfields of Dreams"?: Challenge and Limits of Voluntary Cleanup Programs and Incentives," 1996 *U. Ill. L. Rev.* 883 (1996).

45 Ibid., Eisen at 983 (quoting U.S. Envtl. Protection Agency, Guidance on Settlements with Prospective Purchasers of Contaminated Property). EPA's 1995 Policy for Risk Characterization emphasizes the importance of clarity, transparency, reasonableness and consistency in risk assessments.

46 See 1992 U.S. EPA Exposure Assessment Guidelines; 1992 EPA Risk Assessment Council Guidance; 1995 EPA Policy for Risk Characterization.

47 In 1986, in an effort to settle the debate, Congress amended CERCLA with the Superfund Amendments and Reauthorization Act (SARA). See Pub. L. No. 99-499, 100 Stat. 1613-1781 (codified at 42 U.S.C. §§ 9601-9675 (1994)). The amendments created a presumption in favor of permanent clean-up remedies, while placing a heightened emphasis on the protection of human health and the environment. See ibid. § 9621(b)(1) and (d)(1). For this reason, the EPA generally set the level of clean-up at the residential level, since it was possible that even an industrialized area might be used for residential purposes in the future.

48 See USEPA, EPA/540/1-89/002, Risk Assessment Guidance for Superfund, Volume 1, *Human Health Evaluation Manual* (Part A), Interim Final (1989).

49 See Elliott P. Laws, U.S.E.P.A., Land Use in the CERCLA Remedy Selection Process, OSWER Directive No. 9355.7-04 (May 25, 1995).

50 Such conventional physical barriers may include liners or concrete formation designed to encapsulate or otherwise shield harmful contaminants from human exposure.

51 Specifically, CERCLA requires that all remedial actions satisfy any federal or state environmental law that is either directly applicable or "relevant and appropriate under the circumstances." 42 U.S.C. § 9621(d) (1994). This vague statutory language has created the need to consult a multitude of federal and state regulations to determine the necessary level of cleaning up for a contaminated site.

52 The EPA defines environmental audits as "systematic, documented, periodic and objective review[s] by regulated entities of facility operations and practices related to meeting environmental requirements." *Environmental Auditing Policy Statement*, 51 Fed. Reg. 25,004 (1986).

53 In particular, the EPA policy states that the EPA will "take into account, on a case-by-case basis, the honest and genuine efforts of regulated entities to avoid and promptly correct

violations and . . . [m]ay exercise its discretion to consider such actions as honest and genuine efforts to assure compliance." *Environmental Auditing Policy Statement*, 51 Fed. Reg. 25,004 (1986). However, the EPA does not guarantee that a company's audit will not be used against them, only that the "EPA will not routinely request environmental audit reports." Ibid. at 25,005.

54 As of May 18, 1999, twenty-four states have adopted some version of a privilege and/or penalty immunity law. See USEPA Region 5, Office of Regional Counsel, *Environmental Audits and Self-Disclosures* (modified May 18, 1999) <http://www.epa.gov/grtlakes/orc/audits/audit_apil.htm> In addition, another thirteen states have adopted a self-disclosure policy under which (1) a company's environmental audits will not be routinely requested during an investigation, and (2) voluntary disclosures revealing environmental violations will either not routinely be penalized or will receive a reduced penalty (see ibid.). For example, Massachusetts adopted a self-disclosure policy in 1997, which states that the Massachusetts Department of Environmental Protection will not routinely request environmental audits unless the agency has independent information of a violation. See *Interim Policy on Incentives for Self-Policing: Environmental Audit Policy*, policy Enf-97.004 (visited October 12, 1999) <http://www.state.ma.us/dep/enf97004.pdf> In addition, Massachusetts' policy indicates that companies who satisfy certain criteria will be eligible for a 50 percent reduced monetary penalty and a recommendation to the Attorney-General that no criminal penalties be imposed. See *id.* at 7. The criteria that must be satisfied in order to be eligible for the audit privileges include that the violation be discovered voluntarily, that it be disclosed to the agency within ten days of discovery, and that it be corrected within thirty days of discovery (see ibid. at 10).

55 See 26 U.S.C. § 162 *et. seq.* (1994).

56 Ibid. § 263.

57 See TAM 9315004, 93 TNT 85-16, TAM 9240004, 92 TNT 201-16, and TAM 9411002, 94 TNT 54-13. 26 U.S.C. § 6110(j)(3). While TAMs have no precedential value, they are generally indicative of the direction the IRS will pursue as a matter of policy.

58 See IRS Revenue Ruling 94-38, 1994-1 C.B. 35.

59 Taxpayer Relief Act (HR2014/PL105-34).

60 On January 25, 1996, the IRS issued TAM 9627002, DOC96-19322, which reversed TAM 9541005 and allowed for the current deduction of consulting and legal fees incurred in anticipation of any actual remediation.

61 See Internal Revenue Service Regulations § 1.263(a)–1(b) and § 1.263(a)–2(a).

62 Chapter 206, §§ 34–5, of the Acts of 1998.

63 See Mass. Gen. Laws ch. 59, § 59A.

Bibliography

Bartsch, Charles (1998) Brownfields "State of the State" report: 50-state program roundup. Washington, DC: Northeast – Midwest Institute.

Bartsch, Charles and Collaton, Elizabeth, (1997) Brownfields: Cleaning and Reusing Contaminated Properties. Westport, Connecticut: Praeger Publishers..

Breggin, Linda K. *et al.* (1999) "State Superfund Programs: An Overview of the Environmental Law Institute's (ELI's) 1998 Research," 4 *Alb. L. Envtl. Outlook* 1 (7) (Winter).

Browner, Carol M. (1999) "Statement of Carol M. Browner, Administrator, U.S. Environmental Protection Agency Before the Subcommittee on Water Resources and Environment, U.S. House of Representatives" (May 12, 1999) <http://www.epa.gov/superfund/new/congress/05-12-99.htm> (visited October 6, 1999).

Calland, Dean A. *et al.* (1995) "Analysis and Perspective: Trends in Corporate Environmental Liability: A Comparison Between the United States and Europe," *BNA International Environment Daily* (Aug. 2, 1995).

Croutch, William H. (1997) "Environmental Audits: Should A New Evidentiary Privilege Be Formulated Or Do Existing Privileges Provide Adequate Protection?," 46 *Drake L. Rev.* 425.

Davis, Todd S. and Margolis, Kevin D. (1997) Brownfields: A Comprehensive Guide to Redeveloping

Contaminated Property. American Bar Association, Section on Natural Resources, Energy and Environmental Law (Chicago, Illinois).

Doolittle, Ian (1996) "Environmental Law of the United Kingdom," in *Comparative Environmental Law & Regulation* (Nicholas A. Robinson ed. Dobbs Ferry, New York: Oceana Publications, Inc.) (November).

Eisen, Joel B. (1996) " 'Brownfields of Dreams'?: Challenge and Limits of Voluntary Cleanup Programs and Incentives," 1996 *U. Ill. L. Rev.* 883.

Geltman, Elizabeth Glass (1997) *Prospective Purchaser Agreements, Reducing the Liability Risks of Contaminated Property*, American Bar Association, Section of Natural Resources, Energy and Environmental Law. Chicago, Illinois.

Kubasek, Nancy *et al.* (1998) "Mandatory Environmental Auditing: A Better Way to Secure Environmental Protection in the United States and Canada," 18 *J. Land Resources & Envtl. L.* 261.

Massachusetts Department of Environmental Protection (1996) *Interim Policy on Incentives for Self-Policing: Environmental Audit Policy*, Policy Enf-97.004 (visited October 12, 1999) <http://www.state.ma.us/dep/enf/enf97004.pdf>

Percival, Robert C *et al.* *Environmental Regulation: Law Science & Policy* (3rd edition). Aspen Publishers, Gaithersburg, Maryland.

Posser, Herbert *et al.* (1997) "Environmental Law of Germany," in *Comparative Environmental Law & Regulation* 24 (Nicholas A. Robinson ed., Oceana Publications, Inc.) (April).

Sinisi, Vincenzo *et al.* (1996), "Environmental Law of Italy," in *Comparative Environmental Law & Regulation* 18 (Nicholas A. Robinson ed., Oceana Publications, Inc.) (November).

Trevor Adams College of Law, London (1997) "Environmental Law in the European Communities," in *Comparative Environmental Law & Regulation* (Nicholas A. Robinson ed., Oceana Publications, Inc.) (October).

United States Environmental Protection Agency, *The Brownfields Economic Redevelopment Initiative: Proposal Guidelines for Brownfields Assessment Demonstration Pilots* (visited October 8, 1999) <http://www.epa.gov/swerosps/bf/html-doc/appbook.htm>

United States Environmental Protection Agency (1986) *Environmental Auditing Policy Statement*, 51 Fed. Reg. 25,004.

United States Environmental Protection Agency Region 5, Office of Regional Counsel, *Environmental Audits and Self-Disclosures* (modified May 18, 1999) <http://www.epa.gov/grtlakes/orc/audits/audit_apil.htm>

Wagner, Travis P. (1999) *The Complete Guide to the Hazardous Waste Regulations*. New York: John Wiley & Sons.

Part II
Integrating technology and design

Part II comprises essays and accompanying responses that introduce one of the central themes associated with *Manufactured Sites*: that of the relationship and integration of technology and design within the development of contaminated urban sites. The professional activities of testing, analyzing and remediating contaminated soils and groundwater and the conventional and innovative processes of environmental engineering and construction are described and explored.

In each of the first three essays, the authors, Lucinda Jackson, Eric Carman and Steven Rock discuss the integration of a single remediation technology with future site proposals. The innovative remediation technology, phytoremediation (the use of plants to curtail or uptake soil or water bound contamination), is introduced into the broader sphere of site regeneration and redevelopment planning in three separate ways. Jackson, a scientist and ecologist (Chevron Corporation) argues for clean-up technologies to support integrated site restoration and habitat renewal efforts. Carman, an environmental engineer (ARCADIS Geharty & Miller), argues for a more operational view of innovative technologies. Here innovative technologies contribute to an interdisciplinary engineering and project management approach to manufactured sites as part of a toolbox of clean-up strategies that also incorporate site design strategies. Rock (US.EPA), presenting from the regulatory point of view, offers a cautious approach based on the future potential of phyto-technologies, while admitting the need for continued research into the basic science and applications of the technology. Rock introduces the parallel concerns of landscape architect and engineer in looking at plants from different professional viewpoints.

In the next four essays the engineering of sites is examined. These feature the development of digital tools for assisting brownfield development presented by Sue McNeil and Deborah Lange (The Brownfields Center at CMU and the University of Pittsburgh). An assessment of the extent of site contamination using geostatistical methods by Dante Tedaldi (Bechtel National, Inc.) is followed by an examination of differing attitudes to the integration of interdisciplinary working by Lorna Walker and Richard Owen (Arup Environmental, London). Finally, following discussion of the environmental industries approach to manufactured sites and their remediation, environmental engineer Jean Rogers discusses models for innovative manufacturing and production centers and the implications for future industrial sites.

Chapter 3

Beyond clean-up of manufactured sites:

remediation, restoration and renewal of habitat

Lucinda Jackson

Introduction

Traditional ways of dealing with the clean-up of industrial sites have been to dig up existing contamination and haul it to a hazardous waste dump or entomb it in place with an asphalt or concrete cap. Now we recognize that we can go beyond traditional methods of site clean-up with the use of introduced plants to remediate contamination in place and simultaneously restore and renew habitat. This use of vegetation opens up opportunities for landscape architects, engineers, and biologists to work together to create new environments in former industrial settings.

There are diverse situations within the oil, gas, and chemical industries where site remediation may be desired: along pipelines, at refineries or chemical plants, at former service stations, or at abandoned production sites. These industries are located in various geographical and environmental conditions around the world, from international to domestic, in tropical to arid climates, centralized urban to rural sites, and in industrial to developing countries. In the environmental field, we look for ways to solve remediation issues while concentrating on a policy of pollution prevention, product stewardship, and resource conservation. It makes good business sense to look for solutions that, while leading to clean-up and closure of a site, are cost-effective, environmentally sound, and aesthetically pleasing.

Two case studies are presented to illustrate technologies for remediation, restoration, and wildlife habitat renewal at industrial sites that could incorporate landscape design. Both sites formerly housed industrial businesses: the first case study site was a gasoline transfer terminal; the second an oil refinery.

Case study 1: gasoline transfer terminal

The study site, as shown in Figure 3.1, is a former transfer terminal, where gasoline was transferred from pipelines and tanks to trucks for local shipping. It is a small urban area, about 4 acres, along the main entrance to Ogden, Utah. From the operations that occurred there for over fifty years, accidental gasoline spills and leaks led to current petroleum contamination of soil and groundwater at the site. Traditional estimated clean-up costs at the site are $1 million. In collaboration with the U.S. Environmental Protection Agency (EPA), phytoremediation, the use of green plants to remove, degrade, and stabilize environmental contaminants, was considered as a lower-cost, environmentally friendly remediation approach at the site. An extensive sampling plan was initiated to detect levels

Figure 3.1 Former petroleum terminal site in Ogden, Utah

and location of contaminants. A planting plan consisting of grasses, alfalfa, and trees (poplar and juniper) was designed to locate the vegetation in reference to the contamination areas as shown in Figure 3.2.

A follow-up sampling and monitoring program determined contaminant levels in the soil and groundwater. In three years, all plants were growing vigorously, as shown in Figure 3.3, and sampling results indicated that vegetation, especially grasses, can reduce petroleum contamination in soil.

The mode of action for grasses appears to be enhancement of micro-organisms that degrade contaminants. Poplar trees can inhibit the flow of contaminated groundwater and enhance petroleum degradation. The trees remove water from shallow aquifers and degrade the contaminants trapped on soil particles, with the roots intercepting contaminated water before it can go off-site. The cost of the operation was $200,000, 20 percent of the estimated cost for traditional remediation.

Beyond remediation is the opportunity to combine phytoremediation with site design and planning. As the technology develops, we can learn more about which plants are suitable for site clean-up, including horticultural and native species, and incorporate site design and planning criteria into the phytoremediation planting plan. The benefits of this partnership are increased species diversity (since now phytoremediation is generally done with monocultures of a single plant species or a simple plant mix), aesthetic enhancement, and more functional habitat since their creation could be incorporated into the design process. Challenges exist, including the limited plant species that we know at this time are conducive to phytoremediation, the presence of contaminants that may inhibit plant growth and evoke regulatory issues, and the possibility of human and wildlife interactions with contaminated vegetation (human health and ecological risk assessment).

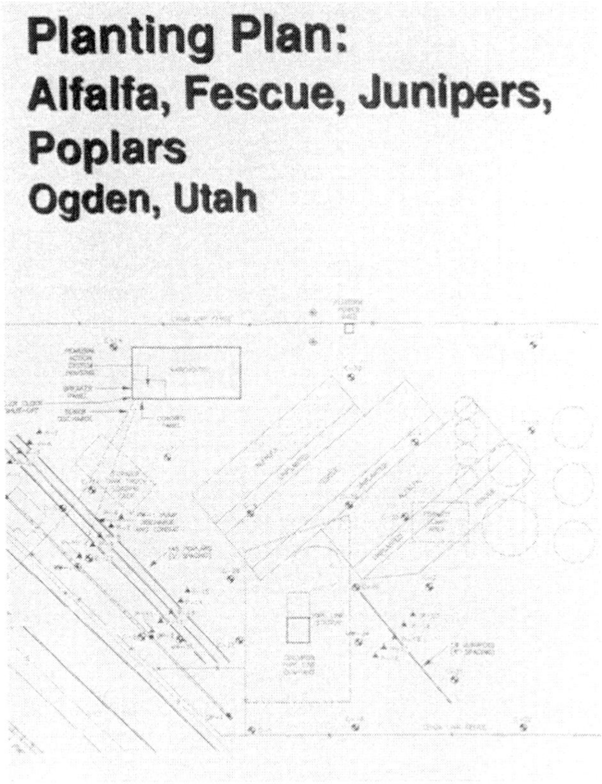

Planting Plan:
Alfalfa, Fescue, Junipers, Poplars
Ogden, Utah

Figure 3.2 *Planting plan to situate plants where they can be most effective in cleaning up contaminated areas*

Figure 3.3 *Vigorous growth of trees, grasses, and alfalfa planted for site clean-up*

Case study 2: oil refinery

The second site is a former oil refinery located on a large, 600-acre site in a rural setting near Cincinnati, Ohio as shown in Figure 3.4. It is situated along the Great Miami River and contains existing habitat for wildlife.

The site has been occupied for over a hundred years by numerous industrial businesses that have operated here and now contains petroleum and metal contamination. Estimated site clean-up is between $50–$150 million, although the site when completely clean will be worth much less (current prices in the area are around $1,000/acre). An important point is that many of these industrial site restorations are expensive processes with no financial return to the seller. Budgets are often tight for amenities such as aesthetic considerations.

The clean-up operation at the oil refinery site is regulated by the U.S. Environmental Protection Agency (EPA). A Community Action Panel, a group of interested local citizens who meet once a month, is also involved. The EPA has final decision-making power over what occurs at the site, and the opinions and direction from the Community Action Panel are also taken into consideration.

At this site, a risk management and land use approach is under consideration. This is a holistic approach that takes into account the expected future use of the land (industrial, recreational, etc.), the desires and concerns of the community, the human health and ecological risks, and the net environmental benefit of any actions. From interviews, area assessments, and real estate analysis a conceptual plan for the future use of the site was developed as shown in Figure 3.5.

The "mixed use scenario" includes a light industrial park, commercial office development, a landscape nursery, an active recreation area (e.g. ball fields), a hiking trail, and a habitat restoration area. Within all of these potential uses, landscape architectural considerations would be applicable, although several issues must be considered that make the incorporation of landscape design somewhat of a challenge.

Figure 3.4 Former petroleum refinery site in Cincinnati, Ohio

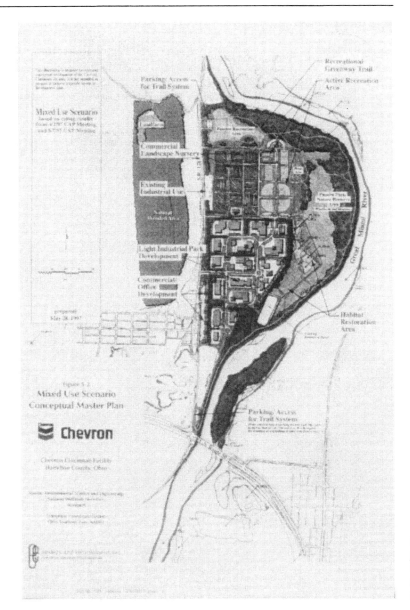

Figure 3.5 Conceptual plan for future land use of petroleum refinery site following remediation

The common issue at industrial sites is that human health and ecological risk must be taken into account. Intensive sampling, conceptual modeling, and testing must occur to identify health risks to humans and wildlife (especially any threatened or endangered species). Both soils and groundwater need to be considered and site-specific analysis must take place. The risks are tied to the future land use – for example, whether the site will be used as a school (higher risk) or industrial site (lower risk). Ecological risk is especially pertinent for landscape architects as it accounts for potential effects of contaminants on plants, animals, and habitat. The tool of ecological risk assessment identifies contaminants and their levels, identifies "receptors" (plants and animals), develops pathway models, determines data needs, and determines if there is a risk to plants and animals and if remediation is needed. If remediation is needed, the next question is how it should be conducted.

One tool that can help with this question is an analysis and decision-making process called net environmental benefit analysis. It balances the benefits of a planned remediation against "ecological costs." It promotes minimal remediation by recognizing ecological sensitivities and potential destructive properties of a remedial action – "the cure being worse than the disease." It also provides for restoration-based compensation. A hypothetical example of net environmental benefit analysis is shown in Figure 3.6.

A traditional remediation such as soil excavation may result in an environmental "debit" due to the destructiveness of the operation, with an overall negative net environmental benefit. If only a partial excavation is considered and, assuming risk is minimal, some contamination is left in place, less disruption of the site ecosystem occurs (less of a "debit"). Cost savings are used to restore a more valuable ecosystem elsewhere (an environmental "credit") and the net environmental benefit of the remediation is positive. Once all this is taken into account and approved by regulators, a landscape architect could be involved in the restoration and site reclamation that would focus on the use of native species and creation of wildlife habitat.

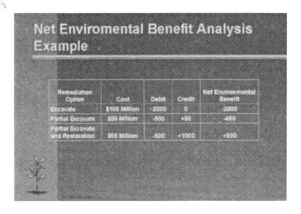

Figure 3.6 Hypothetical example of net environmental benefit of three remediation approaches

Integration of landscape design and manufactured sites

These two case studies illustrate the complexities and benefits of integrating landscape design at manufactured sites. At these sites, we are often dealing with contaminants. There is a need for complete human health and ecological risk assessments at industrial sites before landscape design can be considered. Additionally, with the use of net environmental benefit analysis, landscape design at an industrial site may involve restoration with native species toward the goal of wildlife habitat development. A landscape architect at an industrial site must also then be aware of creating an "attractive nuisance" – an ecosystem that may have some contamination posing a risk to wildlife attracted to the restored site.

Land use scenarios developed for the site must also be considered, and design proposals must be within the constraints of the functional land use determined for the site. Further, there is a need for regulatory agreement for every operation at an agency-regulated industrial site and landscape architects may find themselves in the position of negotiating with government regulators.

Phytoremediation and restoration efforts offer excellent opportunities for biologists, engineers, and landscape architects to work together. New and emerging technologies and approaches give us the chance to create aesthetically enriched, functional environments at former manufactured sites.

Response

Remediation, design, and environmental benefit

Industry: regulations and innovations

The Chevron Corporation (in the two case studies discussed) is exploring nontraditional methods of site remediation. This may reveal a paradigm shift occurring in big business – a shift from mechanically cleaning a site to biologically based solutions. However, as Jackson notes, remediation is undertaken when regulations mandate it, and innovative technologies may be primarily employed when found to be less expensive than traditional methods. Regulations and incentives alone will not create healthy ecosystems – they have to be a priority of a culture's value system.

Phytoremediation

The essay highlights the fact that collecting pertinent site data is critical to a successful remediation. In the urban case study site, a sampling plan was initiated to detect levels and location of contaminants. Follow-up questions to this include: What are the criteria one uses to sample a site? For example, does a consistent grid need to be placed over the site where samples are then taken and mapped? Does one research the history of the site to determine previous land use, vegetation, and groundwater flows? In both of the case studies discussed, Chevron worked with the EPA. Did the EPA set the criteria for sampling a site, the level it must be remediated to, and establish the monitoring program?

Once the contaminants on the site are understood, if phytoremediation is to be employed, one must determine which plants are best for remediating the particular site's soil and/or water. And, more precisely, which plants work best for specific types of contaminants and why. The essay states that grasses, alfalfa, and trees were utilized, but it was not clear if this clean-up effort was designed to be experimental, or simply used available research. The project has been monitored to determine its effectiveness at cleansing soil and groundwater, but one may ask how its effectiveness was measured. Were the plants harvested and toxins extracted and weighed? Was the soil sampled at successive intervals?

The second case study, which is a master plan for a mixed use development, highlights the importance of data collection on contaminants. Planning future land uses is tied to understanding the contaminants present, the clean-up options, and determining necessary levels of remediation for the expected land use. This underscores the need to map and record contaminants and clean-up; as land uses typically change over time, additional remediation could be required for a future land use.

Habitat

What is the relationship between plants for remediating contamination and the creation of habitat? Does one employ phytoremediation for cleansing a site, then when it is clean seek to replace habitat? Once plants are planted and growing, there will be microbial action, insects, and on up the food chain, creating habitat whether intended as such or not. The question of birds and mammals eating contaminated plants was raised in the essay: what is the effect of this in the ecosystem? And what are the potential dangers to human health? It seems clear that more research needs to be done in this area.

The creation of habitat on a site must be seen as part of larger systems of climate, landform, vegetation, and water. It is essential to understand what type of habitat you are trying to restore. Understanding the landscape at a much larger scale than the actual site is critical. For example, some types of bird habitat, and that of large mammals, must take into account many hundreds of acres. The location of nearby wildlife preserves, state forests, lakes, etc. all have a relationship to the fragmented habitat that you may be seeking to renew. These are critical issues far beyond the notion of simply planting native plants.

Remediation, values, and master planning

The second case study site was described as being a hundred years old, and while industry may have been present on the site for a hundred years, the site is of course vastly older than that. It may be simply a choice of words, or it may reveal how differently land is looked at depending on one's point of view. The essay states that remediating this site will cost more than the land will be worth when it is clean. This highlights how a financially oriented value system views land as a commodity, while an ecological value system values the healthy functioning of the ecosystem.

The essay introduces the concept of a "net environmental benefit analysis" which is designed to balance the planned remediation against ecological costs, stating that sometimes minimal remediation is less disruptive to the ecosystem than complete remediation. This minimal remediation saves money that can be invested in restoring a "more valuable ecosystem." What is a "more valuable ecosystem"? More valuable to whom and why? And who makes these decisions?

A master plan for the second case study site has been developed by the EPA, Chevron, and a community action panel of interested local citizens. This approach was described as holistic, yet it is unclear which design professionals, if any, were involved in this plan for the site as a mixed use development. Jackson states, "Within all these potential uses, landscape architectural considerations would be applicable, although several issues must be considered that make the incorporation of landscape design somewhat of a challenge." Rather than being called in at the end of a project to "landscape" the site, landscape architects are the ideal professionals to evaluate the site from a variety of perspectives and effectively design the mixed use scenario (if appropriate) that she describes, and to coordinate with government, business, and citizen groups.

Jackson notes that there are additional implications for designers as more becomes known about phytoremediation and its cleansing properties and aesthetic qualities. Phytoremediation species are typically planted in monocultures, but as more becomes known about which plants clean what contaminants, the designer will be able to choose between (and mix) several plant species, enlarging the range of design solutions. Diversity of plant species also enhances habitat creation. Some plants used for phytoremediation, such as poplar trees, may remain long term on the site, thereby providing not just cleansing benefits but ongoing visual impact.

It is clear that industry, as demonstrated by the Chevron Corporation, is interested in expanding its range of remediation options. Industry and universities, or other research institutions, may benefit by collaborating on experimental sites to determine larger plant palettes for remediation, and to monitor the effectiveness of selected plants – especially as a huge range of sites worldwide may need attention. Including designers in the process from the beginning would provide valuable expertise to the remediation team.

Chapter 4

From laboratory to landscape:

a case history and possible future direction
for phyto-enhanced soil bioremediation

Eric Carman

Introduction

There is an emerging trend in the environmental industry toward identification of the important role of natural processes in mitigating environmental contamination and implementing more passive remedial technologies. As this trend evolves, natural processes will be integrated with more conventional remedial technologies that are commonly used in the industry. This essay presents a broad overview of bioremediation and phytoremediation, two innovative technologies that harness natural processes and can easily be integrated with more conventional remedial technologies. A case history of a phytoremediation project and possible future directions for phytoremediation are also discussed.

Bioremediation is a relatively established remedial technology. It consists of the stimulation of micro-organisms to break down organic contaminants into simpler, often more benign compounds. Generally, micro-organisms that are naturally present in soil and groundwater systems (indigenous micro-organisms) are used in bioremediation applications. Bioremediation has the advantage that it can be used either *ex situ* (above ground) or *in situ* (in place).

Phytoremediation is the use of plants for remediation of soil, sediments, and water. Plants are reported to remediate contaminated environments by several processes, as shown in Figure 4.1.

Phytoremediation processes include either direct uptake of contaminants by plants and the resulting accumulation, biodegradation, or volatilization of those contaminants and enhancement of the biodegradation process in root zone, which is referred to as the rhizosphere. The evapotranspiration processes of plants can also be used to reduce infiltration of surface water, serving as natural barriers or caps.

Certain trees, known as phreatophytes, take up large volumes of water and can be used to control and treat contaminated groundwater plumes hydraulically. At a site in New Jersey, for example, nitrate concentration in groundwater was significantly reduced following implementation of a specially designed phytoremediation program (Gatliff 1994).

There are numerous applications of phytoremediation, and the applications will be refined as the technology matures. It is a technology that holds promise to cost-effectively address sites contaminated with moderately hydrophobic compounds, such as petroleum hydrocarbons, chlorinated solvents, munitions, and excess nutrients (Schnoor *et al.* 1995). In addition, phytoremediation holds promise toward the clean-up of sites contaminated

PHYTOREMEDIATION PROCESSES

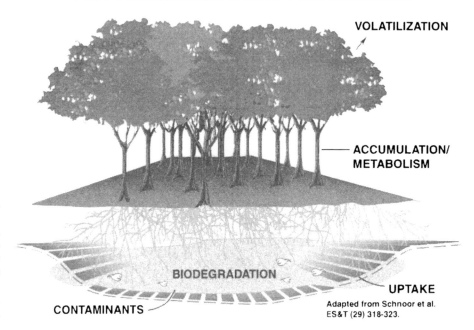

Figure 4.1 Phytoremediation processes include direct uptake, accumulation, biodegradation or volatilization of contaminants, and enhancement of biodegradation processes in the rhizosphere

with heavy metals (Azadpour and Matthews 1996) and recalcitrant organics such as those found at manufactured gas plant (MGP) and refinery sites (Schwab and Banks 1994). Often these are the same types of contaminants that affect abandoned urban sites, commonly known as brownfield sites. Phreatophytes, with their natural capping capacity, have also been used in solid waste applications. Densely planted poplar trees can actually be used as an alternative to conventional landfill caps.

Current applications of phytoremediation include treatment of the vadose zone soil (soil above the water table), shallow groundwater, constructed wetlands and alternative landfill caps. Phytoremediation, like bioremediation, can be used both *ex situ* and *in situ*.

The advantages of phytoremediation are the low cost of the technology (it is generally about one-fifth of the cost of more conventional technologies), environmental compatibility, and public acceptance. Phytoremediation can also be used as a single treatment technology, or it can be coupled with more aggressive conventional technologies such as excavation and treatment/disposal, or less aggressive technologies such as natural attenuation. In addition, phytoremediation can be integrated with landscape design practices such that a remediation system can be a positive addition to the property.

There are, however, disadvantages to phytoremediation that make it unsuitable or undesirable for some environmental applications. Phytoremediation is a long-term remedial technology at most sites, with treatment times on the order of several years. In addition, the technology can be implemented only where the contaminants are present at depths within about twenty feet of the land surface and for contaminants that are located less than a few feet below the water table.

Case history

For a site in Wisconsin, enhanced rhizosphere activity by a species of phreatophyte was employed to stimulate the biodegradation of diesel range organics (DROs) in an aged fuel oil spill. The objective of the project is to remediate soil and fill materials contaminated with DROs within four identified soil hot spots at the facility to below 1,000 mg/kg DROs, as required by the Wisconsin Department of Natural Resources (WDNR). Excavation and treatment of the soil materials was not a preferred option, based on the potential costs associated with segregating construction debris from soil and the risks inherent in excavating at a facility with a long operational history.

Operations began at this site in the early 1900s. A heterogeneous mixture of fill material was used to extend the property boundary west to an adjacent river. During the late 1970s, a section of below-ground piping transferring No. 2 fuel oil failed, resulting in a subsurface release. Approximately 15,000 gallons (56,800 liters) of fuel oil were recovered from shallow trenches installed at the site, and concentrations of hydrocarbon constituents in groundwater are currently below regulatory standards.

Investigations at the site have determined that highly contaminated soil (concentrations greater than 1,000 ppm DROs) remain in four hot spots at the site as shown in Figure 4.2. Three of the hot spots (Hot Spots 1, 2 and 4) are below a hard-packed gravel equipment storage area, and a fourth (Hot Spot 3) is located below a vegetated area along the river.

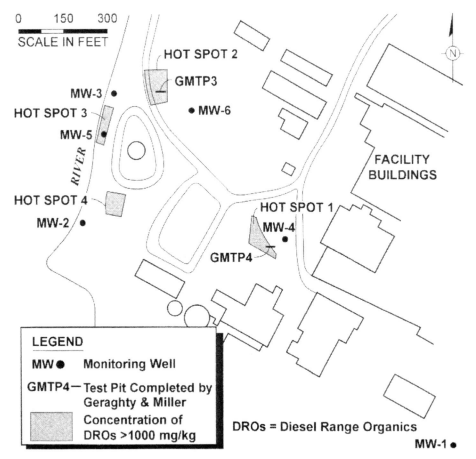

Figure 4.2 Site schematic and locations of the four soil hot spots

Materials and methods

Soil samples were collected and analyzed from 1994 to 1996 to determine initial concentrations of DROs in soil, microbial population densities and respiration rates, and for performing accelerated bioventing tests. Accelerated bioventing tests were performed on two soil samples (Soil Samples GMTP3/4-5 and GMTP4/4-5) for a 24-week period with sacrificing of sample aliquots for DROs or respirometry testing at Weeks 0, 2, 4, 6, 13 and 24. Soil samples were also collected from Hot Spot 2 and Hot Spot 3 to determine phyto-toxic effects on tree root development, concentrations of agronomic constituents of interest, and to select optimal tree species.

Results

Concentrations of the DROs exhibit a wide degree of variability, and the highest concentrations were present in the zone above the shallow water table. Concentrations of DROs ranged from 40 mg/kg to 5,000 mg/kg. The results from the microbial enumerations indicated that a viable population of microbes capable of degrading fuel oil was indigenous to soil at the facility. Populations of heterotrophs ranged from 1.1×10^5 cfu/g to 4.1×10^6 cfu/g and populations of diesel fuel-degrading microbes ranged from 5.7×10^3 cfu/g to 9.5×10^5 cfu/g.

The results of DROs analyses performed on aliquots of Soil Samples GMTP3/4-5 and GMTP4/4-5 are presented in Figure 4.3.

After an initial apparent increase in DRO concentrations between Weeks 2 and 4, an overall 40 to 90 percent decrease in DROs was observed over the course of the 24-week bioventing study. The variability in the concentration of the DROs initially observed over the first several weeks of the study is primarily attributed to sampling variability resulting from the extremely heterogeneous distribution of the petroleum hydrocarbons observed in soil at the site.

Based on the results of this study, *in situ* bioventing and phytoremediation were selected as appropriate treatment alternatives for full-scale site remediation. Phytoremediation was then selected as the alternative, based on its lower costs for implementing and monitoring. Both hybrid poplars and willows exhibited good aerial growth during the root development portion of the agronomic assessment. Willows, however, demonstrated a more pronounced tendency to establish rooting within the DROs-contaminated soil and were selected for planting in the hot spots.

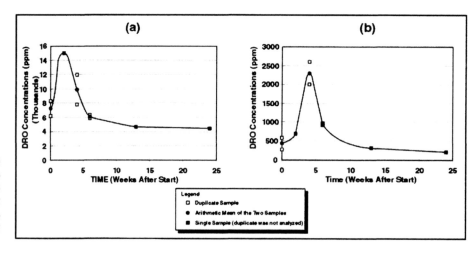

Figure 4.3 DRO concentration over time in biovented soil for (a) Soil Sample GMTP3/4–5 and for (b) Soil Sample GMTP4/4–5

Figure 4.4 Willow tree growth in Hot Spot 2 following the May 1996 planting

Tree planting

Hybrid willow trees (*Prairie cascade*) were planted in the four hot spots during May 1996. The trees were planted roughly at a spacing of 8 feet (2.4 m), with a total of 300 trees planted. Figure 4.4 illustrates the trees in Hot Spot 2 following planting in May 1996.

Trees were planted using a proprietary process which focuses rooting activity and rhizosphere development in the zone of contamination (Gatliff 1994). Site visits were made periodically through the summer months of 1996 to monitor the growth of the trees. Operation and maintenance activities, including plant tissue sampling, fertilizing and insecticide applications have been implemented at the site. Hot Spot 2 in September 1997 is shown in Figure 4.5. Trees across the site grew an average of 4 to 6 feet (1.2 to 1.8 m) in the first four growing seasons of the phytoremediation program. The trees planted in more shallow depths exhibited faster and more vibrant initial growth than those planted deeper. However, following the first two growing seasons, the trees planted deeper exhibited apparent accelerated growth rates. Tree mortality has been significantly lower than expected, and only seven trees, or approximately 2 percent of those that were initially

Figure 4.5 Willow tree growth in Hot Spot 2 in September, 1997

planted have died. During Fall 1997 and 1999, test pits were completed to monitor the rooting activity within the rhizosphere of the trees. In 1997, the roots had yet to penetrate the DRO-impacted soil. However during the 1999 visit, the test pits showed that the tree roots had begun to penetrate the DRO-impacted soils, which is critical to the success of the phytoremediation program.

Soil samples were collected during Fall 1999 for analysis of DROP to aid in determining the progress of the phytoremediation program. Although there was variability in the soil sample results, the concentrations of DRO decreased 66 percent to 68 percent in Hot Spots 2 and 4 and over 50 percent within Hotspot 1 following four growing seasons. Hotspot 3 was inaccessible for sampling during the 1999 field visit.

Future activities

Plant tissue samples will be collected on an annual basis to monitor nutrient concentrations in the trees and to help evaluate the fertilization program, soil sampling is scheduled again for October 2002.

Future trends in phytoremediation

The future for phytoremediation holds promise in the overall remediation marketplace as the trend toward more passive remedial technologies and recognition of the importance of natural processes continues. There will be growth in three areas: the types of vegetation used for phytoremediation, expanded applications of phytoremediation and integration of phytoremediation with natural processes, engineered systems and landscape architectural design and site planning.

Currently the types of vegetation that have been shown to be effective in phytoremediation are relatively limited. As the technology of phytoremediation matures, additional species of plants will be identified for contaminant-specific applications and site-specific applications. Genetically engineered plants also show promise for use on expanded types

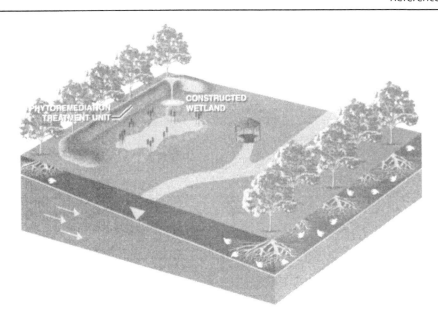

Figure 4.6 A proposed brownfields site in Milwaukee, Wisconsin combined phytoremediation, natural processes, conventional technologies and site design into an innovative remedial approach

of contaminants. Engineered phyto-treatment units (EPTUs, similar to biopiles), constructed wetlands and vegetative caps will be used on a wider range of sites. Phytoremediation will be used increasingly as a remedial alternative for brownfields redevelopment, based on its aesthetic appeal and low cost.

In the future, phytoremediation, natural processes and conventional remedial technologies will be merged through innovative landscape architectural design for an effective, low-cost site remedial strategy. For example, EPTUs, vegetation and constructed wetlands could be integrated with conventional excavation and groundwater extraction for the remedial approach illustrated in Figure 4.6.

Such an approach was proposed at an urban former MGP site with organic contamination of soil and shallow groundwater. This approach would have used excavation and soil redistribution to construct EPTUs, which could enhance biodegradation of organics in soil. Trees planted down-gradient on the property would be coupled with natural attenuation to remove and treat shallow contaminated groundwater on a seasonal basis. More conventional groundwater extraction would supplement the trees on a seasonal basis and water pumped would be treated with a constructed wetland. Vegetation planted on the EPTU and in the constructed wetland would be specifically selected to address contaminants present at the site. The estimated cost for the integrated system was approximately $1,000,000 which is still about 20 percent of soil excavation and thermal treatment, the remedial technologies that were finally selected.

References

Azadpour A. and J.E. Matthews (1996) "Remediation of Metals Contaminated Sites Using Plants," *Remediation*, Summer, pp. 1–18.

Gatliff, G. (1994) "Vegetative Remediation Process Offers Advantages Over Traditional Pump and Treat Technologies," *Remediation*, Summer, pp. 343–52.

Schnoor, J.L., L.A. Licht, S.C. McCutcheon, N.L. Wolfe and L.H. Carreira (1995) "Phytoremediation of Organic and Nutrient Contaminants," *Environmental Science & Technology* 29 (7), pp. 318–23.

Schwab, A.P. and M.K. Banks (1994) "Biologically Mediated Dissipation of Polyaromatic Hydrocarbons in the Root Zone," in T.A. Anderson and J.R. Coats (eds) *Bioremediation Through Rhizosphere Technology*, American Chemical Society, ACS Symposium Series 563, pp. 132–41.

Response

Remediation as engineering?

Arising from a relatively new area of research science, phyto-enhanced soil bioremediation projects may be entering a new phase. In this evolution, the landscape architect and other specialists can play a greater role in developing project goals and design criteria. The results of this collaborative design process may lead to a fresh reinterpretation of neglected contaminated sites for a wide range of users. As an integrated team model evolves and design teams become more diverse, the technology of phyto-enhanced soil bioremediation will have a dialogue with the art of landscape design. Vegetation will be used to cleanse the soil and at the same time to form spatial relationships, mark activity areas and circulation routes, create aesthetic compositions and forms, and provide habitat for wildlife. These technical and design considerations will be strong determinants at the outset when project goals are initially determined. And we may foresee remediation projects that bear a closer resemblance to the landscapes of Capability Brown than to the plantation grid, as so many do today. The design approach for future remediation sites may, in time, go well beyond references to traditional park design, incorporating, as Carman points out, a natural systems approach that not only addresses toxin removal, but creates sustainable new habitats. In combination as well with new water treatment technologies (see Wendi Goldsmith, Chapter 12, this volume), phytoremediation strategies will surely offer more complex and ecologically diverse solutions. The use of natural processes will call for a longer view in the analysis of costs, effectiveness, and benefits, especially when compared to more conventional technologies such as artificial capping, and pumping and drainage collection treatments. In addition to the ecological benefits and long-term lower costs, Carman points to a greater public acceptance of phytoremediation. Presumably "green" technology on a site is viewed as an aesthetically pleasing solution. A note of caution must be raised. If phyto-enhanced soil bioremediation technology is perceived as an aesthetically and psychologically acceptable solution, might we also run the risk of developing a public complacency toward future contamination? In other words, as we treat toxic sites with new "green" technologies do we run the risk of losing the public vigilance against future contamination?

While this essay raises a number of compelling reasons for the diversification of the current technologies and practices through an interdisciplinary approach, a number of questions, both technical and design-related, need further clarification if current and future designers are to embrace, contribute to, and expand upon the obvious potential inherent in this area of practice. As we consider park-like uses and the inclusion of human activity into these sites a number of questions remain as to the degree of environmental and human safety of phytoremediation. As the tree roots absorb toxins in the soil, these toxins would apparently be harbored in the tree. Do dangerous exposures remain possible for children or wildlife in the dead wood and leaf litter? What about pollen and insect life? The migration of these toxins and their persistence in the food web will need to be determined. Are there sites with levels and concentrations of contaminants that exceed the capacity of phytoremediation? In addition, will these projects result in a monoculture of tree species, a culture with minimal diversity, or can a mixed-species approach be developed that offers a rich and diverse ecology? Design considerations would look for the widest possible palette of plant materials, including shrubs and perennials from which to draw and apply effectively in remediation projects. These questions will clearly require further research and the answers will be critical if the uses of these projects are to be expanded.

While cost-effectiveness and efficiency will remain driving determinants in addressing disturbed sites, the potential for multiple and long-term benefits including parkland, wildlife sanctuaries and greenbelts will win strong advocacy. In the early stages these projects should become design laboratories, where process and results are evaluated with data created for further refinement and integration in subsequent projects. Environmental designers will find the greatest potential in the combination of various models of constructed wetlands, phytoremediation and conventional remediation technologies, integrated in a multi-modeled approach to site remediation. From the evidence suggested in this essay, it seems clear that future efforts in the area of phyto-enhanced soil remediation will follow the model of integrated design teams, teams that will include scientists, designers, ecologists and other specialists. How these projects, in process and product, will differ from the traditional engineered projects is not clear.

While the scientists, as evidenced during this conference, are eager to engage the designers, designers will need to become more knowledgeable of the latest technologies, so they can bring informed proposals to the table. Clearly this dialogue will need to take part at many levels, and as the disciplines gather around the table, scientists must educate designers and designers must engage scientists about the creative process. As the complexion of the teams changes, so too will the working process as programming and aesthetic and spatial considerations are integrated into the planning.

Chapter 5

Phytoremediation:

integrating art and engineering through planting

Steven Rock

Introduction

Landscape Architecture and Remediation Engineering are related fields, united by common areas of endeavor, yet they have strikingly different languages, techniques, and habits of thought. A professional from each discipline may stand at the same place, view the same site, and describe the site with different words that express a vastly different experience and professional perspective. While the landscape architect may interpret particular qualities of the land shaping the site through its contours and artifacts, the remediation engineer is concerned with groundwater flows and soil contaminant concentrations, human health and the environment.

What unites the fields is the fact that they often work on the same site, with the common goal of making what may be a derelict and hazardous landscape once again functionally and aesthetically integrated. Both professions use some of the same tools, for example GIS mapping to evaluate a location, or earthmovers to control run-off and shape the site. What separates the professions is time. Typically, remediation is complete before the landscape architect is brought onto the project. Projects may benefit from a closer integration of these often disparate fields, and not all sites need to be completely remediated before reuse, especially if part of the reuse helps clean the site.

Introducing phytoremediation

Phytoremediation is the use of living planted material to clean environmental hazards. Work is underway to develop and evaluate planted systems that remove or detoxify contaminants from soil, sediments, and groundwater as shown in Figure 5.1.

On some sites it is possible to place planting in such a way as to allow partial reuse of the site for public access or ongoing development while the clean-up is in process. Here phytoremediation and creative site design are united by the use of planted systems that both remediate and at the same time establish spatial and functional patterns of use.

When a plant encounters a contaminant, in soil, water or air, it has several possible interactions: it may find the contaminant toxic and die, it may ignore it completely, it may transform the contaminant into products that are useful to the plant, or it may take the element into the root or shoot of the plant and store it. Phytoremediation is a set of technologies that harness those natural processes in order to clean up a site that has pollutants that are hazardous to humans and the rest of the environment.

Figure 5.1 Cottonwood trees in a greenhouse treatability study, Cincinnati, Ohio

In order effectively to use plants as engineering tools, those natural plant processes or mechanisms must be scrutinized and understood, a mindset that I call "learning to think like a tree." Trees, animals, bacteria all have some chemicals that, while potentially harmful, are safe or are even needed by them for other organisms. In order to learn to think like a tree, or bacteria, it is useful to imagine that the room in which you are sitting is suddenly and completely filled with jelly beans, and the only way for you to survive and escape is to eat your way out to the nearest door. You could do it if four basic conditions were met: (1) you had sufficient air; (2) you had sufficient water; (3) you had an additional food supply (man, nor bacteria, cannot live on jelly beans alone); and (4) you could avoid poisoning yourself on your own wastes. It may take a while, but given these conditions it would be possible to consume vast quantities of jelly beans, something that normally a person will eat in limited quantities, if at all.

So it is with plants. Most plants prefer clean air, water and soil, but out of the 200,000 plant species many have evolved opportunistic adaptations that allow individual species to thrive in sub-optimal conditions. Some plants accumulate metals, some degrade toxic compounds and use the resultant nitrogen or carbon, and some enable bacteria to thrive in the root zone creating synergies known as the rhizosphere effect.

Phyto mechanisms

The mechanisms of phytoremediation can be grouped into three broad categories. These categories are based on the fate of the target contaminant, and include accumulation, degradation, and hydraulic control. The following sections describe each of those mechanisms and illustrate each with either a field application or a case study.

Accumulation

There are several species of plants that will take heavy metals into plant tissue. Plants may store the metals in the roots. Sunflowers accumulate large quantities of uranium in their root structure. A floating growth platform in the Ukraine and a hydroponic greenhouse system in Ohio have shown that these plants can accumulate metals in roots. Poplar trees can take metals into the space between root cells, and these intracellular contaminant concentrations can exceed the expected toxic levels for the plant. Certain plants have been identified that not only take in metals through the root structure, but then translocate the accumulated metals from the root to the leaf and shoot. While many plants do this function to some extent, some plants, known as hyperaccumulators, can concentrate as much as several parts per hundred, contaminants to dry weight of plant. The plant can then be harvested, and the heavy metals recovered or disposed of, cleaning the soil of the contamination.

Often the plants that naturally hyperaccumulate are native to remote and unique locations, and are not suitable to the temperate climates which are found in some parts of the US. Some of the hyperaccumulators are small in size (the Indian mustard plants shown in Figure 5. 2, for example), so that even though they accumulate significant percentages of heavy metals, their biomass in so low that the total amount of contaminant removed from the ground is low.

To overcome these limitations, standard or slightly modified crops and agriculture practices are being tried, with Indian mustard being used as a plant that can both accumulate significant quantities of metals and be grown easily in many parts of this country. Chemical amendments to the soil may be used to mobilize the metal into the plant.

In Trenton, New Jersey, a battery manufacturing facility was discovered to have contaminated the ground around the building with lead. The cost of removing the soil was prohibitive, especially since the original polluter was bankrupt. A phytoextraction demonstration is underway with Phytotech Inc. of Monmouth Junction, New Jersey, and the US EPA SITE program to evaluate the efficacy of the technology. This two-year study will evaluate the uptake of lead by Indian mustard and sunflower crops. The soil on this site, and many other similar brownfields, is more like a gravel parking lot than a garden in texture, composition and structure. First it is tilled and agricultural amendments are added. The crop is planted either by broadcast or in rows. Weed control and irrigation are decided on a site by site basis. As the plant matures amendments are added to alter the soil

Figure 5.2 Indian mustard plants for metals uptake, Findlay, Ohio

chemistry to allow the metal to be readily transported into the plant. Shortly after that round of amendments are added, the plants are harvested. The crop is air dried and sampled for metal content. If the metal content is high enough the metal-laden plant can be sent to a smelter for metals recovery. In the greenhouse these metals have reached 2 percent of dried plant mass, which is a higher concentration than some conventionally mined ores. If recycling is not an option, the material can be ashed and sent to a hazardous waste landfill, at a considerable saving to taking contaminated soil to the landfill.

Degradation

This is the second category of phytoremediation, and it can also occur in two parts of the plant system: the root zone or rhizosphere, or in the plant tissue directly. Plants have tremendous beneficial effects on the microbial population in the rhizosphere. As a result of the symbiosis that occurs between a plant and its microbial neighbors, microbial populations have been reported to be two orders of magnitude higher in the soil of the root zone than in adjacent unplanted soil. The roots of many plants provide the growth-limiting factors: oxygen, nutrients, and water to the microbial zone as a by-product of normal plant growth. As roots penetrate the soil, there is passive aeration as the roots loosen the soil, and active aeration as the roots release oxygen as part of normal plant respiration. Parts of tree roots die off during seasonal water and temperature fluctuations. Abandoned or sloughed roots and root hairs become a nutrient source to the rhizosphere microbial community. These nutrients may serve as cometabolites, sustaining microbes that incidentally degrade contamination as part of their metabolism. Plants also enhance microbial communities by drawing water into the near surface root zone. Dropping leaves and sloughing roots adds to the organic matter content of the soil, which adds to the soil's ability to retain water.

Plants may degrade contamination during metabolism in the plant, transforming it to a less toxic form. Various plants produce different enzymes, many of which are useful in the destruction of contaminants. Nitroreductase removes nitrogen from TNT, dehalogenase can pull chlorine from a chlorinated solvent. These enzymes and others can either detoxify a contaminant directly, or render it vulnerable to microbial consumption.

The USA has many sites contaminated by organic compounds from oil refinery spills, pipeline leaks, old municipal gas plants, and leaking storage tanks. The EPA and Phytokinetics Inc. from Ogden, Utah conducted tests at a former wood-treating site where railroad ties and telephone poles were treated with creosote. Over time creosote leaked into the soil. Annual ryegrass was planted and compared to biodegradation for two growing seasons. The grass was planted with soil amendments, seeded and irrigated to optimum agricultural levels. The plants outperformed the controls by reaching some of the clean-up levels faster, though the non-planted treatments achieved nearly the same level over time.

Hydraulic control

This includes the influence plants can have on the movement of contaminants, especially in groundwater. Poplar trees, as shown in Figure 5.3, as well as cottonwood and the willow family, have been shown to draw as much as 350 gallons of water per day. Water usage for a two-year-old tree in the same climate is more like 3–5 gallons per day. Water consumption at even that rate can lower an aquifer level, preventing the spread of contaminated groundwater. Lowering the aquifer level may also prevent contact between a relatively clean groundwater and a more contaminated shallow soil, perhaps in concert with an extraction procedure. Volatile contaminants may be taken up with the water and volatilized through normal tree transpiration. Another physical effect that plants can have on contaminated sites is dust and erosion control, through the formation of a vegetative cap. A cap consisting of grasses, clovers, shrubs and trees is being used to prevent both

Figure 5.3 Fast-growing hybrid poplar trees (three years old), Odgen, Utah

wind and water erosion on a Montana mining site. Caps consisting of trees and grasses have been used to cap municipal landfills, and are being proposed across the country.

EPA and the US Air Force (USAF) have initiated a field demonstration designed to evaluate the effectiveness of eastern cottonwood trees in remediating shallow groundwater contaminated with trichloroethene (TCE). Phytoremediation of groundwater involves planting deep-rooted, water-loving vegetation to reduce contaminant levels in the saturated zone. The demonstration entails planting and cultivating eastern cottonwood trees over a dissolved TCE plume in a shallow (6 to 11 feet below grade) alluvial aquifer.

The cottonwood trees are expected to bioremediate the contaminated groundwater and any contaminated soil through one or more of the following mechanisms:

- Release of root exudates and enzymes stimulating microbial activity in the rhizosphere and enhancing biochemical transformations of contaminants.
- Metabolism or mineralization of contaminants within the vegetative tissues; the contaminated water enters the vegetative tissues by root uptake from the aquifer.
- Transpiration of water by the leaves.

In essence, the trees are expected to serve as a natural pump-and-treat system.

TCE concentrations in the groundwater, soil from the rhizosphere, and tree tissue will also be monitored during the demonstration. In general, data will be gathered and interpreted to identify the overall effect of the planted trees on the dissolved TCE plume in the aquifer. Changes in the flow field and the position of the TCE plume will also be modeled. Microbial activity in the rhizosphere will be monitored and transpiration rates will be measured. These data will be used to determine the fate of the TCE at the site, including those processes that affect its fate.

Tree plantations or groves may be used to remediate shallow groundwater and soil contaminated with TCE, as well as other contaminants common to USAF installations. Such contaminants include petroleum, munitions, and halogenated hydrocarbons. Costs of the technology are limited to initial site preparation, planting, and occasional maintenance (irrigation). Phytoremediation can be integrated with landscape architecture, as shown in Figure 5.4.

At present the typical process for a contaminated site, either Superfund or brownfield, is site characterization, remediation plan, remediation, then site and reuse design including infrastructure and implementation. Using phytoremediation could change that paradigm. A site may use trees for long-term containment of a groundwater plume at the same time as utilizing the shade and framing functions of the trees for the above-ground site design. Finding a new use for an old tool is the essence of innovative technology; finding a dual use is even better. The application of plants, one of the essential tools of landscape architecture, to the engineering pursuit of site remediation may require involving both professions at an early stage in the site discussion so that the eventual use of the site is decided early enough so that the plants can be installed and remain undisturbed.

Plants may be described as self-assembling, solar powered, pump-and-treat units, but they are also living organisms that have strengths and limitation. They are subject to animal consumption, insect infestation, disease, fire, and other events that limit their

Figure 5.4 Trees intercepting and cleaning groundwater while providing shade for people and habitat for wildlife, Connecticut

effectiveness. They also change their work during non-daylight hours, and shut down entirely during certain times of the year. On the positive side they are sometimes self-repairing, and seem to encourage less human vandalism than mechanical systems. A great deal of research remains before the mechanisms of phytoremediation are sufficiently understood for widespread application of the various technologies.

For a remediation engineer plants may seem less reliable and predictable than physical, mechanical, or chemical systems. For the landscape architect, trying to incorporate remediation goals in a planting plan may be frustrating through limits to planting palette and the placement of vegetation. For the public, as long as the health risk is equally averted by all remediation options, phytoremediation enjoys a certain intuitive support. Green growing sites feel environmentally benign. For the ecosystem, planting a site can be a first step to habitat creation, soil enhancement, erosion control, and other benefits, alongside contaminant reduction and progressive design.

Response

Recovery with plants

The healing properties of plants have been understood over the ages. From the ancient medicinal use of herbs and roots to the current use of urban canopy street trees to filter dust and absorb air pollutants, plants have been selectively used as tools to promote environmental and human health.

"Finding a new use for an old tool is the essence of innovative technology; finding a dual use is even better." Rock's statement in his introduction to the science, technology and application of phytoremediation illustrates his experience of integrating technology and site design through the application of plants, or more correctly introduced planting systems on manufactured sites.

Some of the broader issues currently facing manufactured sites include treating the social malaise that accompanies these sites to create livable neighborhoods, the development of environmental education about urban reclamation and manufactured sites, and assistance in advancing trends in technology, particularly regarding innovations in site assessment and remediation.

The use of phytoremediation as an innovative method of clean-up, as outlined by Rock, directly addresses a number of these issues – from the ability to connect planting systems used for clean-up, to larger site open space and vegetation strategies, to the use of phyto-technologies within community greening programs for inner-city environments.

There are two issues related to the use of phyto-technologies with site design and planning strategies that are significant for future interdisciplinary work between engineers and designers. These are: (1) the differing concerns of time between remediation and development practices; (2) what may be termed the difference in mindsets between phyto-engineers and site designers regarding the forms of planted systems and their ongoing performance, operation and predictability. In short, the expectations for the duration and site application of phytoremediation

Time

A phytoremediation system can be thought of as part of the delivery of a usable site and construction program in one of two ways. First, over the long term (say thirty years), where phytoremediation is "embedded" in evolving interim and temporary land-use and site programs, or second, where anticipated development is in an accelerated time-frame of nine months to three years. Here phytoremediation systems are implemented as part of the delivery of a usable site and construction program.

Application

Phytoremediation plants perform in different ways and are viewed in a quite different light from the standard conventions of horticultural and landscape design. In reviewing the literature of phytoremediation, one is struck by the insistence in the scientific and industry-research papers on viewing the planted systems as derived from an agricultural scale of application (the use of terms such as "crop," "fields," "harvest" and "hedgerows").

There are a number of issues that still have to be addressed before remediation and site design can be truly integrated. Receiving attention from industry researchers and regulatory agencies, such as the US.EPA National Risk Management Laboratory in Cincinnatti where Rock is based, are such topics as plant selection and matching to contaminant type, the use of native plants and habitat restoration in phytoremediation, the evaluation of claims, and the continued exploration of plant and contaminant mechanisms.

Phytoremediation is at a point in its evolution where the nature, scale, complexity and location of manufactured sites can start to shape the application possibilities for smaller urban sites. Among these are the existing urban site conditions of ground and water, the plant growth concerns on these sites from microclimate and poor soils, and the concerns of adjacent communities and stakeholders.

Chapter 6

Engineering urban brownfield development:

examples from Pittsburgh

Sue McNeil and Deborah Lange

> Brownfields are abandoned, idled or underused industrial or commercial facilities where expansion or redevelopment is complicated by real or perceived environmental contamination.
>
> (Brownfields definition, US.EPA 1996)

Introduction

Engineering urban brownfield development has a broader meaning, as the revitalization of these sites must address a diverse set of issues. These include community attitudes, environmental issues, the provision of new or the renewal of existing infrastructure, urban and land use planning, socio-economic issues, historical legacies, financial issues, legal and regulatory environment and political forces. This essay looks at each of these issues and explores some of the contributions research has made to facilitating this process. Two case study sites from Pittsburgh are used to illustrate the concepts presented in this chapter.

Diverse perspectives on brownfield development

The diverse perspectives on brownfield development add complexity to the problem as stakeholders struggle to determine alternatives that are feasible, and ideally, optimal. The stakeholders include owners; developers, communities, economic development agencies, city, regional, county, and state governments, and regulators.

Community attitude

Communities are critical stakeholders who significantly influence the set of acceptable alternatives or options. Community support can "make or break" a project, but rarely do the affected parties interact at the early stages of development. The most successful brownfield developments have involved a bottom-up effort that involved the community in seeking alternative site uses and maintaining the momentum in the rather lengthy development process. For example, positive community involvement was instrumental in finding an alternative use for a former munitions factory in Hays, a Pittsburgh community (Barton 1996).

Environmental

Environmental regulations and liability are generally regarded as the single most significant deterrent to site redevelopment (Zhang 1998). However, as states adopt Brownfields legislation, such as Act 2 in Pennsylvania (Land Recycling 1995) that sets clean-up levels to be consistent with intended use, environmental issues have taken a back seat as the technical problems are generally solvable with sufficient time and money.

Identifying and documenting the historical uses of the site can play an important role in the identification of the source of the contamination and the determination of an appropriate remediation strategy.

Infrastructure

The presence and condition of infrastructure at, or adjacent to, a site influences site specific uses, the speed of the negotiation, and later the development process. Infrastructure includes roads and railways, waterways, water/sewage and utilities. In Pittsburgh, many brownfields were industrial sites prior to the highway era, therefore road access is often limited. However, river and rail access is likely to be available and functional. The condition of infrastructure is an important item for negotiation and many innovative financing strategies focus on this issue.

The debate over the relationship between economic growth and development and the provision of infrastructure (Gramlich 1994) provides no clear answers as to whether or not one drives the other. Either way, it is clear that they are closely related. Sites cannot function without adequate supporting infrastructure.

Urban/land-use planning

Urban brownfields range in size from abandoned gas stations to industrial giants of many thousands of acres. The larger tracts can influence urban form and the visual landscape. Where the identification of appropriate site uses is a challenge, designing developments that are consistent with good principles of urban design is a relatively straightforward task. Implementing these designs is another issue that must recognize funding realities and the need to integrate the redeveloped site with the surrounding areas. For example, the development of Washington's Landing and the proposed development of Nine Mile Run in Pittsburgh both include tracts of urban residential space that are based on principles of neo-traditional neighborhoods (Putaro and Weisbrod 1997; "Ample Opportunity" 1998).

Planning also includes the preservation and development of greenspace and the principles of sustainability. Bringing together interdisciplinary teams that include artists, engineers, designers and planners in the context of brownfield development provides an opportunity to try to define the role greenspace plays and to identify the aspects of these projects that contribute to sustainable development (Collins and Savage 1998).

Socio-economic

The specific uses of a site are heavily influenced by demographics. This includes the population density, the educational backgrounds, the age distribution, and the racial diversity of the population in the surrounding area. Potential developers and site end-users will evaluate the available labor force, the potential market, the social service priorities, and the support services available in the surrounding neighborhoods, communities and region.

Historic

A historical perspective on a specific site provides considerable insight into the environmental legacy left by previous occupants. For example, development of the Pittsburgh

Technology Center was halted for several years as engineers and scientists were confounded by the presence of ferrous cyanide. Review of the Sanborn maps determined that a manufactured gas plant had occupied the site prior to the steel mill. The cyanide, generated from the purification of manufactured gas, was found to be localized and stable, and then building was able to proceed (Messenger and Santoro 1996).

History also provides significant insight into the social evolution of a site, its proximity to transportation and neighboring residential areas, and the concept of site criticality as specific sites serve as the focus for a community or a common thread that holds the community together. Although the majority of an industry's employees may not be drawn from the adjacent community, the industrial focus was often the thread that united the community. When the industry leaves, the community loses its hope and vitality.

Financial

The financing of brownfield developments is one of the more complex issues facing the stakeholders. A wide range of sources of money is available. Often particular sources are earmarked for particular activities. For example, clean-up monies can come from a variety of different sources, infrastructure improvements can be paid for with tax increment financing or make use of state infrastructure banks. Clearly, the most successful developments have been public/private partnerships where many stakeholders have a vested interest in the success of the development.

Legal/regulatory

The legal and regulatory environment (at the state and federal levels) has had a significant impact on the phenomenon of brownfields. State and federal legislation has allowed for clean-up consistent with intended use, and limited liability of lenders. Questions still exist in terms of responsibilities for pre-existing conditions, future discoveries and future regulations. Pennsylvania Acts 1 and 2 provided protection from some of the liability issues and paved the way for responsible clean-up consistent with the intended use of the site (Land Recycling 1995).

Other legal and regulatory issues include compliance with zoning, stormwater drainage, driveway access and transportation impacts.

Political

Brownfields can represent a significant asset to the communities in which they sit; therefore, political issues can be difficult. The existence of political support for these issues relates to the overall economy, the power of urban versus suburban voters and the nature of the area.

An interdisciplinary approach to brownfield issues

The Brownfields Center (TBC), a joint venture of the Carnegie Mellon University and the University of Pittsburgh, focuses on research, education and outreach, and as such it serves as a link between academia and the communities.

Faculty, staff and students participating in this interdisciplinary center come from varied backgrounds including art, architecture, economics, decision sciences, civil and environmental engineering, history, urban planning, economic development and public policy. TBC has also developed links to the STUDIO for Creative Inquiry, and the Center for Economic Development at Carnegie Mellon. TBC also has extensive interactions with federal, state and local government agencies such as Pennsylvania Department of Transportation

(PennDOT), Southwestern Pennsylvania Regional Planning Commission (SPRPC), Urban Redevelopment Authority (URA), City Planning, Pennsylvania Department of Environmental Protection (PaDEP), Environmental Protection Agency (EPA), Housing and Urban Development (HUD), Department of Energy (DOE) and private sector firms such as consultants, banks and developers.

TBC's activities include maintenance of an extensive website with case studies, other brownfield links and reports (http://www.ce.cmu.edu/Brownfields/), participation in conferences as program chairs, session organizers, presenters and exhibitors, and presentation of brownfield issues to diverse audiences. TBC is also actively involved in the training of the workforce through the education of both graduate and undergraduate students who participate in the projects and activities.

The interest in abandoned industrial site redevelopment, and the close ties between urban revitalization and infrastructure renewal, motivated the formation of TBC. These concepts form the foundation for a National Science Foundation (NSF) funded project titled "Brownfield Development: The Implications for Urban Infrastructure." This project builds on research and development related to civil infrastructure systems. It also leverages Carnegie Mellon's location in a city with abundant brownfields and in a state with legislation to encourage brownfield development.

The objective of the NSF project is to develop a rational, consistent and systematic approach to decision-making related to infrastructure renewal through:

- development and transfer of computer-based decision-making tools;
- exploration of the role of computer-based collaborative environments;
- collection of, access to, and management of relevant data;
- analysis of qualitative and quantitative information, site and infrastructure assessments and development of rational decision-making strategies;
- recognition of the historical process and context of industrial site development and redevelopment;
- exploration of financing opportunities.

The project has involved information gathering, exploration of existing tools, tool and model development, system integration and documentation, and technology transfer. Specific components of the project include:

- understanding the brownfield development process through case studies;
- exploration of impact assessment tools;
- enhancement of traffic and transportation models;
- integration of condition assessment and decision-making methodology;
- exploration of life cycle cost analysis.

Some information management tools have also been developed (Amekudzi *et al.* 1998). A Regulation Broker provides access to regulations and goes beyond text searching. An information management system known as RISES (Regional Industrial Site Evaluation System) is built in a geographic information system to provide access to site and related information such as demographics, environmental conditions and economics. For example, Figure 6.1 shows typical demographic information for a site.

Most importantly, TBC provides a link between academia and communities. To illustrate these concepts case studies of two sites are presented. The first, Washington's Landing, is a mixed use development where development is complete. The second, Nine Mile Run, is under development, and the STUDIO for Creative Inquiry has played a critical role in facilitating the public dialogue during the planning process (Collins and Savage 1998; "Ample Opportunity" 1998).

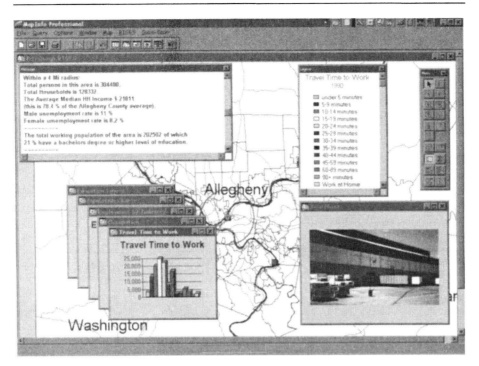

*Figure 6.1
Demographic module in
RISES: typical
information*

Washington's Landing

This site, formerly known as Herr's Island, was renamed to reflect the fact that George Washington slept here (Putaro and Weisbrod 1997). The 60-acre island in the Allegheny River is less than one mile from downtown Pittsburgh, and was originally a rail stop for livestock. As a result, related industries, such as meatpacking and rendering, grew. As early as 1959, the site was proposed for recreation but redevelopment did not start until the late 1980s. The mixed-use site was developed using public/private financing that covered bridge demolition, and infrastructure improvements such as a new bridge to connect the island and East Ohio Street. PCB contaminated soil was relocated to a state-of-the-practice landfill on the site, and rendering wastes and carcasses were disposed off-site. The on-site disposal of the PCB contaminated soil cost approximately $3.4 million. The off-site disposal of the organic wastes cost approximately $800,000. The new developments include tennis courts on the landfill, a 44,000 square foot PaDEP building, townhouses and other recreational facilities. Figure 6.2 illustrates the townhouses and Figure 6.3 the PaDEP building.

As of March 2000 the site is nearing full development and full occupation. The Urban Redevelopment Authority (URA) of Pittsburgh spearheaded the development process with a total public and private investment of approximately $71 million (Stikkers and Tarr 2000).

Nine Mile Run

Nine Mile Run, a former slag dump, is the largest piece of undeveloped property in Pittsburgh. It is 238 acres between the Monongahela River to the south and Frick Park to the

Figure 6.2 Washington's Landing townhouses

Figure 6.3 Washington's Landing office buildings

Figure 6.4 Nine Mile Run

north. To the east and west, the tract abuts existing residential neighborhoods. Figure 6.4 shows part of the site.

URA purchased the site for $3.8 million in 1995. Upscale housing is planned for the area, while maintaining about a hundred acres of greenspace. As plans for the development have proceeded several issues have arisen related to:

- the traffic generated at the site,
- contamination of the stream from leaking sewer systems in neighboring communities,
- concern with airborne contaminants from the earthmoving during site development.

The STUDIO for Creative Inquiry has conducted community workshops and public meetings to work toward a reasonable set of development alternatives that justify the expenditure of approximately $45 million for infrastructure improvements with a total projected public and private investment of $185 million.

Summary

The many diverse perspectives that must be addressed to ensure responsible and desirable brownfield development appear as different layers of complexity in the development process. Pittsburgh provides a rich environment for the study of brownfields that allows us to explore the role of interdisciplinary activities, partnerships between academic communities and government agencies. Underlying our experiences is the critical need for open communication.

Acknowledgements

This work is partially supported by the National Science Foundation under grant CMS-9526029.

References

Amekudzi, Adjo, Paul Fischbeck, James Garrett, Jr, Haris Koutsopoulos, Sue McNeil and Mitchell Small (1998) "Computer Tools to Facilitate Brownfield Development," *Public Works Management and Policy* 2 (3) (January).

"Ample Opportunity: A Community Dialogue" (1998) Final Report, Nine Mile Run Greenway Project, Studio for Creative Inquiry, Carnegie Mellon University, Pittsburgh, Pa. (February).

Barton, J.P. (1996) "Hays" Project Report, Department of Civil and Environmental Engineering, Carnegie Mellon University, Pittsburgh, Pa. (http://www.ce.cmu.edu/Brownfields/NSF/sites/hays/)

Collins, Tim and Kirk Savage (1998) "Brownfields as Places," *Public Works Management and Policy* 2 (3) (January).

Gramlich, Edward (1994) "Infrastructure Investment: A Review Essay," *Journal of Economic Literature*, vol. XXXII (September).

Land Recycling and Environmental Remediation Act 2, The General Assembly of Pennsylvania, Senate Bill No. 1 Session of 1995 (http://www.libertynet.org:80/~pahouse/laws95/95act002.html)

Messenger, Adrienne and Carla Santoro (1996) "Pittsburgh Technology Center," Project Report, Department of Civil and Environmental Engineering, Carnegie Mellon University, Pittsburgh, Pa. (http://www.ce.cmu.edu/Brownfields/NSF/sites/ptc/)

Putaro, Sarah and Kathryn Weisbrod (1997) "Washington's Landing," Project Report, The Brownfields Center, Carnegie Mellon University, Pittsburgh, Pa. (http://www.ce.cmu.edu/Brownfields/NSF/sites/Washland/index.htm)

Stikkers, David E. and Joel A. Tarr (2000) "Pittsburgh's Brownfields: Case Studies in Redevelopment," Working Paper, The Brownfields Center, Carnegie Mellon University, Pittsburgh, Pa. (March).

United States Environmental Protection Agency, Region 5 Office of Public Affairs, Brownfields Fact Sheet, 1996.

Zhang, Jianyu (1998) "Concerns and Priorities of Stakeholders in Brownfields Decision Making," Working Paper, Department of Engineering and Public Policy, Carnegie Mellon University, Pittsburgh, Pa.

Response

Living laboratories: studies in infrastructure and industrial land

Since most American industrial facilities of the eighteenth and nineteenth centuries were located in urban centers, the primary sources for what are now known as "manufactured sites" or brownfields are located in or near our metropolitan centers. The location and diversity of interests woven into these sites and the complex treatments required for clean-up necessitate that strong links be developed between urban design, science, economics, legal, community and public process to develop strategies for successful remediation. Both of the case studies – Washington's Landing and Nine Mile Run – offer tangible examples of how several of these and other issues have been addressed through practice.

As a model for data gathering, information dissemination, and project documentation The Brownfields Center (TBC) at Carnegie Mellon University offers a unique framework to link the interdisciplinary work of academic teams to the concerns of local communities, federal, state and local agencies, private consultants and industry. This alliance suggests a greater function for academics through their traditional role as data collectors, coordinators, researchers and policy advocates, and their institutions can serve as data resource banks, disseminating timely research results and post-construction evaluations of recent remediation projects.

Universities are experienced in working across disciplines and academics can bring their expertise in mediation and community facilitation to the process of brownfield development. TBC offers a practical model integrating the new technologies (the web) with traditional research (case studies) to inform and link diverse participants together and provides a forum in which brownfield issues are discussed and the latest innovations in technology are presented. In addition, web-based accessibility offers a public forum through which debate and discussion take place and both experts and novices can decimate and acquire information.

This model also suggests a framework in which to develop future academic curricula focusing on brownfields, and other forms of environmental remediation. As a visionary center, it offers a practical real world model of interdisciplinary collaboration among students in design, science, law, government and public policy. How might this model be applied at other institutions?

In the model developed by TBC over the past three years the focus has been on urban brownfield remediation. It is not clear if this interest is the result of their location in Pittsburgh, Pennsylvania, a city with abundant brownfield sites, or if it is tied to the vigorous housing and commercial climate in which alternative urban sites are now being considered for development. It may also be because of public policy revisions to change the scope of liability carried by developers and establishing more realistic standards and time-frames for clean-up. Concurrently, and not being directly addressed by the TBC, there is a growing trend to remediate sites outside of urban boundaries. These include abandoned strip or pit mines, landfills, riparian corridors and hazardous storage or manufacturing facilities. While the location of these sites may be rural, the diverse range of issues, including varied stakeholders, socio-economic implications, financial and regulatory complexities, are not dissimilar from those found in the urban context. TBC's model may offer some strategies for addressing similar issues at rural sites. While the political structure and nature of community concerns may differ, many of the technical issues and similar team/partnership models may be applied when addressing rural site problems, including water quality, soil contamination, riparian restoration, landfill closures and innovative "green" stormwater treatment systems. As the type of sites requiring remediation continue to expand, it becomes clear that a model similar to that developed by McNeil and Lange may be equally valuable for these projects.

As McNeil and Lange have pointed out, it is critical to understand the history of the site in order to evaluate the patterns and degree of contamination, since the past historic uses and patterns of contamination have a direct correlation to the design solution. The distribution and selection of uses are in direct response to the levels of contamination and the degree to which the various areas within the site can be cleaned. As is pointed out in Chapter 7 by Dr Dante Tedaldi in the "Tacoma Asarco Smelter Site," the distribution of land uses as shown in the master plan was based on developing uses compatible with the level of clean-up that could be reasonably achieved. In areas that could be substantially cleaned, housing, commercial and office uses were planned. For those where substantial clean-up could not be achieved, a large park was designed with a containment system installed below and a capping system above.

What would be equally interesting, and maybe achievable through the case studies being compiled by TBC, is to see if there are discrepancies between the chosen uses and the level of clean-up achieved per differing communities. In other words are some communities held hostage by past land uses, are differing standards being applied, and is political leverage influencing allowable types of uses and the integrity of the design resolutions?

Since brownfields are defined by their current condition, the sites often vary dramatically in scale. Are some sites too small for development and better suited as laboratories for exploration of alternative remediation methods, as was suggested in Chapter 4 by Eric Carman in "From Laboratory to Landscape"? An example was presented where it was not economically feasible to develop most decommissioned gas stations. In Chapter 3, the essay "Beyond Clean-up of Manufactured Sites: Remediation, Restoration and Renewal of Habitat" by Dr Lucinda Jackson, a value was found in which the sites were considered laboratories for experimentation and data collection. On the larger scale many of these large parcels or linked parcels offer an opportunity for urban renewal and are candidates for integrated open spaces that can be linked to existing open space systems and seen as anchors for economic revitalization. Another strategy to address is smaller peripheral sites that might link prime development sites in the downtown core with contaminated sites in outlying residential areas. Through various incentives and public support a developer could redevelop a downtown site for maximum profit if they agree to take on an outlying site for conversion into housing or residential uses. Remediation should not be a limitation to good urban design.

McNeil and Lange also raise the issue of sustainability. We must be careful in a zest to remediate these sites not to trade in one problem for another. Despite what may be increased costs are we recreating productive landscapes, or are we treating contaminants on-site so that we don't create additional hot spots off-site? Are we using design to create community amenities that re-establish community in a sustainable manner? The challenge we're facing as we move into the twenty-first century is not only to remediate contamination, but to create thoughtfully designed healthy communities that overcome the environmental ravages of our history.

While the diverse perspectives common to these projects, including those of the community, the developer, the politician and banker suggest a complex process, they also offer tremendous opportunities. The inclusionary process proposed in this chapter requires considerable time investment over a long period; however, when successful, it offers a diverse set of potential solutions and innovative strategies, ideas that may not surface in a more constrictive process. Given the great diversity of the stakeholders and varying levels of expertise among the participants, what other models can be borrowed or created to develop clear design criteria, evaluate options, and to maximize economic, environmental and social opportunities? How does a manufactured site model such as that used by Peter Latz (Chapter 2), with a strong focus on community participation and stewardship, compare with models currently being used in the United States? How does the culture and differing political systems found in other countries influence the process? Given the importance of a bottom-up effort with the community why are there still so many projects that lack a front-end outreach component? Clearly the skill level of each community organization will vary depending on their political status, experience with outreach processes and internal cohesiveness. Understanding the abilities of each unique community early in the process will be critical in any bottom-up, inclusive planning. What are some of the best methods to inform the community and solicit their participation without alarming them and beginning the process with suspicion and distrust? What tools can be used to understand the dynamics of the community if in fact it is not organized or clearly identifiable? Who will speak for the community? Should there be multiple voices or is it feasible to develop consensus despite the fact that expectations among the community may vary widely? As the number of case studies compiled increases and methodologies of practice emerge, perhaps some of these questions will be addressed.

The authors also raise several questions, two of which have significant application for future researchers. Are there other successful interdisciplinary models, linking the public with governmental agencies, academic institutes or departments, and private development corporations, that may offer alternate approaches applicable to brownfield remediation? The second question is how can a balance be struck between an open exchange of information, a timely decision-making process, economic feasibility and inclusionary processes so they can operate as complementary components?

Chapter 7

The Tacoma Asarco smelter site:

the use of geostatistics to guide residential soil clean-up

Dante Tedaldi

Introduction

The Ruston/North Tacoma Operable Unit of the Commencement Bay Nearshore/Tide Flats Superfund Site encompasses over 950 acres of almost completely residential land with about 4,300 residents, less than an hour's drive from Seattle, Washington. This formerly blighted, post-industrial site is undergoing remediation that will facilitate public uses through the establishment of parklands and commercial development.

The principal health threats at the site are from ingestion of arsenic and lead contaminated soil, resulting from fugitive emissions from an adjacent former smelter. Several hundred soil samples were collected from the site and analyzed for metals, leachability and other soil characteristics related to contaminant fate processes. Geostatistical analyses were conducted to generate isopleths (a line on a map connecting points at which a given variable has a specified constant value) of lead and arsenic concentrations in soil. The results of the investigation indicated that levels of arsenic and lead were greatly enriched over background levels. Generally, the concentrations of arsenic, lead, and other priority pollutant metals decreased with depth; however, in some areas the depth of contamination extended deeper than 30 cm. Soil contaminated by arsenic and lead above risk-based action levels (230 mg/kg and 500 mg/kg, respectively) is being removed, followed by backfilling with clean soil and on-site disposal of contaminated soil in a secure facility. This massive clean-up of almost 1,000 residential lots is expected to be complete by 2004 at a cost of over $50m. An innovative approach to on-site soil disposal and containment will permit light industrial development and park use.

Site history

In 1983 the area now known as the Commencement Bay Nearshore/Tide Flats site was added to the United States Environmental Protection Agency's (US EPA) list of Superfund sites. The Ruston/North Tacoma Residential Study Area is an operable unit of this large Superfund site and it encompasses several hundred hectares of primarily medium density residential land with over 1,800 homes, as shown in Figure 7.1.

The inclusion of the residential study area was based primarily on soil contamination resulting from atmospheric emissions from an adjacent former smelter, owned and operated by ASARCO Incorporated (Asarco).

The Asarco smelter is located in both the town of Ruston and the city of Tacoma, Wash-

Figure 7.1 Air photograph of site

ington. The smelter property encompasses approximately 80 acres with an adjacent break-water of 23 acres. The smelter began operations in 1890 as a lead smelter and produced 5 tons of lead per day. Asarco, then named the American Smelting and Refining Company, purchased the smelter in 1905 and converted it to a copper smelter in 1912. The smelter had a brick smoke stack, approximately 571 feet tall, which was built in 1917 using 2.5 million bricks. This stack, the source of a significant fraction of the fugitive emissions, was one of the tallest smelter stacks in the world until its destruction in 1993. The smelter specialized in processing ores from around the globe with high arsenic concentrations and recovered arsenic trioxide and metallic arsenic as by-products. Peak production during the early 1980s reached 500 tons of copper per day. Copper smelting operations ceased in 1985 due to changes in the economics of domestic smelting, and the arsenic production plant was closed in 1986.

Site investigations

Early investigatory work was performed at the site in 1988 for the Washington State Department of Ecology. Over the preceding twenty-five years, numerous other human health and soil investigations were conducted in the study area. These investigations led researchers to the conclusion that fugitive emissions from the smelter stack, as well as wind-blow particulates from uncovered ore piles on the site, resulted in the soil

**SURFACE SOIL ARSENIC
CONCENTRATION VERSUS DISTANCE**

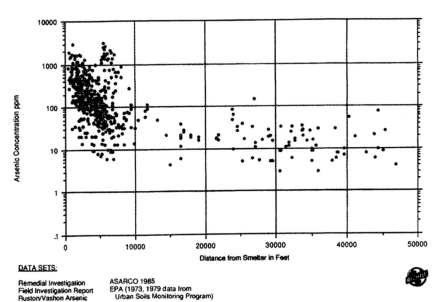

*Figure 7.2 Contaminant
concentration vs
distance graph*

DATA SETS:

Remedial Investigation
Field Investigation Report
Ruston/Vashon Arsenic
    Exposure Pathways Study

ASARCO 1985
EPA (1973, 1979 data from
    Urban Soils Monitoring Program)

contamination now present nearby. The principal contaminants were arsenic and lead, with much lower levels of smelter-related metals such as antimony, copper, cadmium, mercury, and silver. Data illustrated in Figure 7.2 show a clear trend of decreasing concentration in soil with both increasing depth as well as increasing distance from the smelter.

Based on the results of previous investigations a remedial investigation (RI) approach that focused on soils as the primary medium of environmental concern and arsenic as the primary contaminant of concern was developed. Soil sampling and analyses were performed as part of the RI to provide additional information to fill data gaps on the distribution of arsenic and other metals in surface and subsurface soils in the study area. Previous soil data collected by others and new data collected from the RI were combined into a single, large data set. The multi-dimensional nature of the distribution of contaminants over a very large area confounded common and simple interpretive approaches. In addition, the project team desired a data evaluation method that would enable them to estimate contaminant distributions with a defined degree of mathematical certainty. For these reasons the data were analyzed using geostatistical techniques.

For the RI, a total of 222 soil samples were collected from 163 locations throughout the site. Surface and subsurface soil sampling at a limited number of locations provided data on the vertical distribution of arsenic and lead within the first several centimeters, at the 15- to 25-cm depth and the 30- to 41-cm depth. Subsurface sampling locations were concentrated in four directions radiating from the Asarco smelter stack to facilitate an understanding of the effect of distance on contaminant distributions and accumulation.

For all samples collected, total arsenic concentration ranged from a maximum of 3,000 mg/kg to a minimum of 2 mg/kg. These high levels of soil arsenic at the site sharply contrasted with typical background levels. Urban and suburban arsenic concentrations in lawns nationwide were reported to have mean values less than 20 mg/kg and, generally, lead values were well below 400 mg/kg.

Geostatistical analyses

Two different mathematical techniques, nonlinear interpolation and Kriging analyses were used to model and evaluate the spatial patterns of soil contamination in terms of distance, depth and direction. The bivariate quintic interpolation method divides the area defined by the data into a number of triangular cells, and the value at each point in a triangle is estimated (interpolated) by a bivariate fifth degree polynomial. The resulting interpolated surface passes through all the given data points. Interpolation maps do not average out or smooth over any of the raw data points. The resulting maps are complex, reflecting the spatial variability of the data set. Due to this variability these maps do not necessarily highlight the overall patterns in the data – a prime consideration when making general conclusions related to the distribution patterns of hundreds of data points collected over an area of almost a thousand acres.

To compensate for this effect, a second method (Kriging analysis) was also examined and the results were noticeably improved. Kriging, originally developed in South Africa for gold prospecting, assesses the variability among the data values as a function of the distance between the data pairs and then uses this information to estimate concentrations at locations throughout the study area. Each estimated concentration is a weighted average of values from nearby points with weights assigned according to the variance associated with the inter-sample distances. Kriging produces a smoothed surface in which some of the spatial variability of the data set is eliminated. The Kriging results are useful for assessing the overall patterns in the data, but with the loss of some of the fine-grained detail that is actually present in the data set. Anomalous values may be "averaged out" and may not be visible on the Kriging map.

For this large-scale project, data smoothing was an acceptable trade-off that substantially improved the interpretability of the resulting figures, as shown in Figure 7.3. Kriging and interpolation maps were used to define zones of contamination and estimate the size of the areas within the study that exceeded the primary action levels for arsenic and lead.

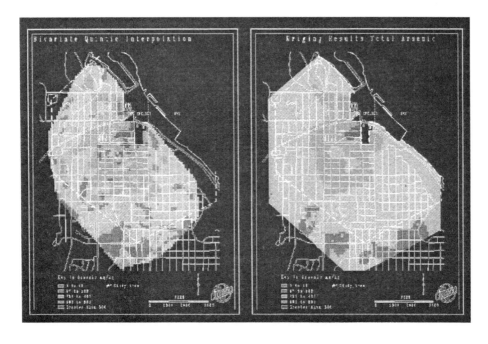

Figure 7.3 Quintic vs Kriging results

Study results

The results of the RI confirmed that soil contamination in the study area is largely the result of emissions from the Asarco smelter. Conventional and Kriging analyses demonstrated that the overall distribution and concentration of arsenic, lead and other smelter signature metals decreases with distance from the stack. In addition, there was a statistically significant correlation between soil concentrations of arsenic and lead, antimony, copper, cadmium, mercury, and silver. This fortuitous finding facilitated clean-up operations because removal of the primary contaminant, arsenic, would generally subsume soil affected by trace levels of secondary contaminants. However, the high density of residential sampling locations indicated that concentrations could be highly variable over short distances, e.g., across a single backyard or a playfield. This fact would tend to increase the complexity of soil removal and clean-up operations. The area where impacts from the smelter can no longer be detected was outside the study area.

The results of the Kriging analysis indicated that approximately 270 acres containing about 525 homes would require remedial action. The properties included in this "action area" were likely to, but not in all cases, have arsenic or lead contamination in concentrations above EPA's action level. The Kriging predictions of a real extent and vertical distribution were directly applied to the clean-up plan development efforts. Excavation quantities, disposal options, and time to complete the clean-up were all a function of the Kriging results. Although the Kriging results were well founded, additional sampling has been performed during the implementation of the remedial action to determine specifically the properties and portions thereof that require remedial action. This is typical and in no way reflective of a diminishment in the value of the Kriging exercise.

The selected remediation alternative

After extensive analysis of many potential alternatives for site clean-up, EPA and the community selected a remedy that involves excavation of soils from properties that exceed EPA's action levels for arsenic and lead. Excavated areas are filled with clean soil and replanted. A key component of the program is concern for the individual homeowner's existing plant material and garden artifacts, and every reasonable effort to restore affected properties to their original condition has been made, as shown in Figure 7. 4.

In general, most property owners have been pleased with the restoration and many properties post-remediation are more aesthetically attractive than they were prior to the clean-up.

Under the current clean-up plan, each property identified in the Kriging analysis is likely to exceed action levels if sampled. Other locations identified as "clean" within the Kriging analysis are sampled on an as-needed basis as requested by property owners. Prior to completion of the clean-up, up to 1,600 residential properties may need to be assessed for further action. Typically, most residential lots have been sampled for arsenic and lead. An individual lot is divided into 5–10 areas of roughly equivalent size; each of these areas is then randomly sampled at four locations and at four depths (0–3 cm, 3–15 cm, 15–31 cm and 31–46 cm). The results of the four random samples are averaged for each depth interval and these values are used for comparisons with the EPA action levels. If any depth interval for a particular sub-area indicates that contamination exceeds action levels, then that sub-area is excavated.

The majority of the properties requiring clean-up can be cleaned completely (i.e., soil above the action levels will not remain), thus eliminating the need for long-term institutional controls on many private properties. Soils that exceed action levels below 46 cm are not excavated. Where contamination exists below this depth, 46 cm of clean soil is substituted for the top 46 cm of native soil and a permanent maintenance and monitoring

Figure 7.4 Remediation sequence

program for the capped areas is established. This program ensures that if contaminated soils are excavated in the future for development or other reasons, proper safety procedures will be followed. Excavated soil is disposed of on the former smelter site and provides a substantial benefit as a recycled material in that it functions as a stable and native fill for the redeveloped area. A significant cost savings is realized through the use of these soils on-site rather than disposal of the soils at an appropriate off-site facility. It is estimated that up to 187,000 cubic yards of soil may require excavation. The process of sampling, testing, excavation, soil replacement and yard restoration is expected to take seven years from the start of operations. The present value (in 1993) of the complete clean-up is approximately $56m, with an average cost per affected residential property of $58,000. Clean-up was initiated in 1994 and has progressed at a rate of 125 clean-ups per year. By 1998 over 450 clean-ups had been completed, with 250 of the tested properties not requiring any form of remediation. Clean-up is expected to be complete by December 2003, and parkland on the former smelter site underlain by excavated contaminated residential soils will be operational by the summer of 2004.

Site redevelopment

Approximately 100 acres of the property comprising the former Asarco smelter site are included in a definitive agreement that was approved in 1996 by Asarco, the town of Ruston, city of Tacoma and the Metropolitan Park District of Tacoma. The master plan is the community's vision for the site, based on many weeks of public workshops. The complex plan simultaneously addresses the need for a comprehensive remediation

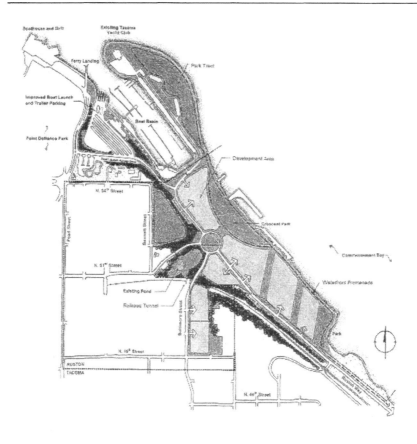

ASARCO MASTER DEVELOPMENT PLAN
PLAN DEFINITION REPORT

*Figure 7.5 Master
development plan*

SCHEME: G 2.1:
BENNETT STREET PROMONTORY

program on the site and the community's interest in future site uses. This plan is a "broad-brush" look at what the site may look like when developers come forward and develop the property in the future.

Of the 100 acres, approximately 37 acres will be appropriate for redevelopment after remediation is completed. The development plan includes new roads, a pedestrian prome-nade along the waterfront, parks, Tacoma Yacht Club facilities, development areas and marine structures, as shown in Figure 7.5.

The new roadway plan includes a roundabout in the center of the former smelter site and this will act as the central focus of the plan, providing two lanes of one-way traffic. Street lighting, sidewalks with wheelchair ramps and marked bicycle lanes are also com-ponents. Parkland will be established directly over buried waste that has been separated from the topsoil by a specified system of protective barrier layers. Hillside plantings will be used to recreate a green face on steep, barren slopes. Parking areas and area lighting will be constructed following site grading and the installation of drainage and irrigation systems. As shown in Figure 7.6, a pedestrian promenade that is accessible for persons with disabilities will include a spacious public walkway linking the existing Ruston Way walkway with Point Defiance Park.

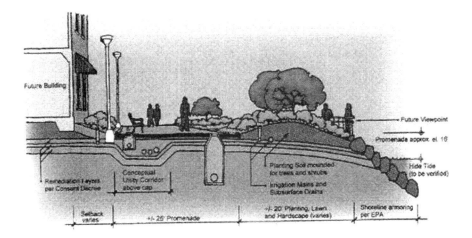

Figure 7.6 Pedestrian promenade

The Tacoma Yacht Club is situated on a man-made peninsula constructed entirely of slag from the former smelter. This 16-acre peninsula will be covered with soil, graded and planted for parkland use. Marine structures, including a boat launch facility and a break-water marina, will also be constructed. The wide range of possible development includes public parks, offices, mixed use, even light industrial facilities. Interest by developers and the marketplace will determine the use of specific parts of the site. The Preservation and Development Authority, a separate public corporation responsible for managing the site, will solicit and evaluate specific proposals from developers for projects such as offices, restaurants, or light industrial facilities. View corridors to scenic Commencement Bay will be respected. The legally enforceable consent decree between the stakeholders specifies that remediation be complete by December 2003. The parks will most likely be available for public use by summer 2004.

Conclusions

The definition of the areas affected by smelter releases was of considerable interest to researchers involved in this project. Due to the high costs associated with repetitive soil sampling and analytical analysis, it was necessary to identify a defensible method of data presentation and assessment that would enable conclusions to be formulated with a relatively small data set. The use of geostatistics, Kriging in particular, proved extremely useful and offered noticeable advantages over more conventional and simplistic forms of data analysis. Although the final data set was large, it was relatively small when one considers the total surface area under study. The use of geostatistical and Kriging analysis enabled the project team to reach clear conclusions about the vertical and lateral extent of contamination. Moreover, when the graphic output of the geostatistical analyses was combined with a geographic information system, the resultant multicolor illustrations provided a convenient and easily understandable presentation format. These illustrations were a critical component of not only technical discussions among consultants, regulatory officials and the responsible parties but also for presentations of summary-level information to the concerned public. At most public meetings, active and involved citizens took great interest in poring over large versions of the maps, looking for the exact location of their home and its relation to the Kriged contaminant contours. As noted earlier, the smoothing effect of the Kriging analysis substantially reduced the number of discrete, visually anomalous areas of concern; i.e., individual grid cells well separated from the contiguous areas of contamination. The smoothing over of these areas, and thus the disappearance of seemingly inexplicable "hot spots," had the significant effect of minimizing citizen concerns for data anomalies.

Another interesting feature of the geostatistical analysis was the ability to assess value that is added through the collection of large numbers of additional soil samples. Thus, one could mathematically determine the increased statistical level of confidence in the resultant data projections attributable to increases in the overall data set. Surprisingly, it was discovered that the recent addition of several hundred samples during the RI did not appreciably increase the level of confidence in data projections. This was attributed to the fact that a large portion of the old and new data were collected close to the smelter source; thus this area was "data heavy" while those areas farther out, along the fringes of the area of concern, were somewhat deficient in data.

The project illustrated the limited value in the collection of excessively large environmental data sets. Specifically, the Kriging analysis demonstrated that carefully selected yet small data sets could be used to develop relatively accurate interpretations of site conditions. This is of particular importance to preliminary studies that will be followed by additional sampling during and after clean-up. Significant cost savings can be achieved through the elimination of redundant sampling during the various phases of a project. This redundancy is often requested by regulatory officials and residents, and becomes especially pronounced during the clean-up and closure phases of a remediation project.

Bibliography

Carey, A.E., J.A. Gowen, T.J. Forehand, T. Han and G.B. Wiersma (1980) "Heavy Metal Concentrations in Soils of Five United States Cities, 1972 Urban Soils Monitoring Program," *Pesticides Monitoring Journal*, 13 (4).

United States Geological Survey (1984) "Element Concentrations in Soils and Other Surficial Materials in the Conterminous United States," H.T. Shacklette and J.G. Boerngen, Professional Paper 1270.

United States Environmental Protection Agency (1988) "GEO-EAS (Geostatistical Environmental Assessment Software Users Guide)," E. Englund and A. Sparks. EPA/600/4/033, Office of Research and Development, Environmental Monitoring Systems Laboratory, Las Vegas, Nev.

Response

Manufacturing site information

As Steven Rock pointed out in Chapter 5, it is possible to take an environmental engineer and a designer who both have a professional interest in the same section of contaminated land and for them to perceive the opportunities and constraints in completely different ways. This not only includes the demarcation and scope of the work itself, but the type of knowledge that each seeks and demands as part of the record of site conditions and information.

In Tedaldi's essay the use of geostatistics is explained as a method of developing, as he states, a "defensible method of data presentation and assessment that would enable conclusions to be formulated with a relatively small data set." This essay should be of extreme interest to planners and designers in explaining an engineering method of using environmental site information as a working tool. This is true both in terms of its use in public forums and as a means of assessing existing information and the ability of new information to render additional value or confidence to the conclusions. For those designers used to the normal conventions of topographical surveys, photographic records of vegetation, soil maps and site sketches, this form of data manipulation may appear bewildering at first – how does a map become "smoothed," where do the contamination "hot-spots" go?

The design professions continue to address complex sites in other fields of endeavor. Surveying radically altering terrain through contour models and spot elevations, or the subtle three-dimensional changes in vegetation patterns in hedgerows or woodlands, requires the adaptation or development of working tools in digital media and diagrammatic models. What makes Tedaldi's descriptions of special interest is the method of translating site knowledge concerning the extent of contaminated soil in three dimensions into usable and readable flat maps and plans.

Chapter 8

Regeneration:

vision, courage and patience

Lorna Walker and Richard Owen

Introduction

This essay describes two successful regeneration projects carried out in Great Britain. Key issues common to both of these projects were the early involvement of the full design team, the provision of public and private financial support, and clients who had vision, courage and recognition that this type of project takes time to mature. It can also be shown that these regeneration projects were a trigger for further development. The two projects present very different characteristics and problems: one is a derelict and disused dock area in Salford in the north-west of England, and the other is a closed landfill site near Heathrow airport, 25 miles from the center of London.

Salford Quays, Manchester

Site location

Salford Quays lies 3 kilometers from the center of Manchester, in Lancashire. Road communications are good, the national highway network can be reached in five minutes. The site is located in an area which, in the mid-1980s, showed all the effects of urban decay and loss of confidence. Nearby is the Trafford Park Estate, once the largest industrial campus in Britain before becoming virtually derelict. To the east there is a mix of post-war municipal housing and industry.

   The site is triangular in shape, formed by the convergence of the ship canal and the Trafford Road, and with the southern boundary of the Enterprise Zone providing the base (see Figure 8.1). The total area is 60 hectares, of which one-third is water. The dock basins range from 256 metres to 823 metres in length, and from 68.5 metres to 76 metres in width. The site is flat, and at the time of redevelopment was almost devoid of any structures of merit apart from the dock walls themselves and a disused swing bridge. The major asset was the presence of water which, conversely, was the source of two major problems. First, it was highly polluted; second, it separated the developable port land into long unconnected fingers.

Site history

In the last quarter of the nineteenth century, industry and commerce were moving from Manchester toward Liverpool to take advantage of the port facilities. The Manchester Ship

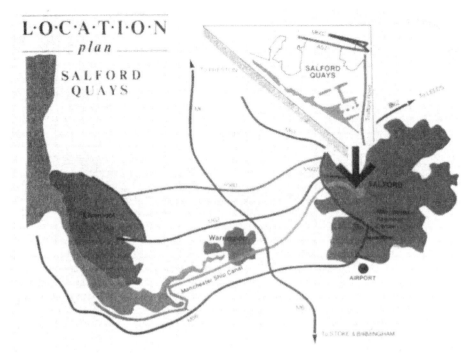

Figure 8.1 Site plan

Canal and port facilities were constructed to respond to this challenge. The Manchester Docks were opened in 1894 following the completion of the Manchester Ship Canal, which runs 35 miles from Eastham on the Mersey Estuary to Salford. This canal enabled sea-going vessels of up 12,500 tonnes to sail into the heart of the conurbation.

Trade peaked in 1958 when 18 million tonnes were handled. However, economic changes increased the size of ships, and failure to respond rapidly to new handling techniques led to a steady decline, culminating in the closure of Salford Docks in 1984.

By the late 1970s, inner Salford had been designated an Inner City Partnership Area under the Inner Urban Area Act 1978, jointly with inner Manchester, and it qualified to receive Urban Programme and Derelict Land funding through the UK government's urban policies. In 1981 the Salford/Trafford Enterprise Zone was designated and encompassed some 150 hectares in Salford, the majority being vacant land owned by the Manchester Ship Canal Company and held for port expansion purposes.

Several attempts were made to promote development around the dock; few developers were interested, however. The City Council proposed developing the area through a series of packages involving different developers. The Council was able to purchase the majority of the dock in late 1983. An overall development plan was commissioned in the hope that it would reassure investors about the long-term future of the docks. This plan would both clarify and guide future development and enable the proposed reclamation works to have a program and a context. Shepard Epstein and Hunter, architects and town planners, were commissioned by the City Council to prepare the plan. Ove Arup and Partners were appointed consulting engineers.

Joint vision for site reclamation

The plan was published in May 1985. It established a reclamation strategy and proposed that the reclamation program should concentrate on the three most important aspects of the site. These were identified as:

- the dock water;
- roads and services;
- public access, landscape and amenities.

Water quality

It was recognized that the dock water was a major asset of the site, although at that time it was highly polluted and considered a major disincentive to potential developers.

The visual appearance of the docks was characterized by oily, dark-colored, foul-smelling water with a large accumulation of floating debris. The docks contained up to 2 meters of sediment derived from fluvial deposits. The heavy metal concentrations of the sediment were low, although plant nutrient concentrations were high and there was a considerable oxygen demand from slowly degrading matter.

Samples were taken and analyzed to characterize the quality of the water. The main problems were identified as high bacterial levels coupled with high nutrient levels. There was very low dissolved oxygen, and thermal stratification occurred during the summer months. These factors led to the formation of eutrophic conditions.

The approach to solving the water quality problems

An approach had to be incorporated into the development plan to include features that satisfied the requirements of both water quality and the development as a whole. The main aspects of the plan, which were influenced by the need to significantly improve the water quality, were:

- the construction of bunds;
- the provision of a double lock gate entrance;
- the linking of water bodies;
- the isolation of the docks from the canal;
- equipment;
- the re-routing of surface water drainage.

The new construction works that were undertaken are illustrated in Figure 8.2.

Bunds or embankments were formed at the end of three of the canal basins. These were created by dredging the silt down to the original clay layer and placing crushed concrete, which was derived from demolition material from the site and adjacent areas. The bunds effectively isolate the basins from the main source of pollution, the Manchester Ship Canal, and allow treatment of the impounded water. They also unite the individual piers and allow a more efficient infrastructure network to be established.

The provision of a double lock entrance connects the basins to the Salford Quays development. The existing basins were linked by the construction of canals. This increases the effective volume of water, which increases tolerance to pollution. The canals also provide an added dimension to the otherwise flat sites and increase the valuable water frontage.

In the early assessments of the dock water it was thought the enclosed basins would be particularly susceptible to eutrophication. The main controlling factor of eutrophication is the availability of the plant nutrients nitrogen and phosphorus. The major source of these nutrients and their loading rates were identified. In reality the only source identified which

Figure 8.2 The new construction works

could be directly controlled was surface water run-off. It was therefore decided to exclude the surface run-off from the enclosed basin by diverting flows either directly into the Manchester Ship Canal or into the existing sewerage network.

Even with the exclusion of stormwater, the inputs of nitrogen and phosphorus still classified the impounded water as dangerously susceptible to eutrophication. Therefore additional measures were required to increase the tolerance of the dock system to eutrophication. These additional measures were identified as artificial mixing systems. These systems eliminate the adverse effects of stratification and reduce those associated with high nutrient loadings, as follows:

- Thermal and chemical stratification is prevented, and oxygen rich water from the surface is distributed down through the full depth of the water column to the sediment.
- The accumulation of algae in the brightly lit surface layer is reduced and algae are forced to spend more time at depth, where the light is limited and their growth rate is reduced.
- The amplitude of diurnal cycles of dissolved oxygen concentrations and pH are reduced and herbivorous zooplankton, under reduced stresses, can survive throughout the water body exerting substantial grazing control on the algae.
- The improved supply of oxygen to the sediment maintains an oxidized layer, which in turn reduces nutrient recycling and sustains an active zoobenthos capable of efficiently processing organic debris settling from the planktonic community above.

The artificial mixing system installed consists of sixteen helixors and three compressors to supply compressed air. Each helixor is an open-ended plastic tube with a helix formed through it. The helixors work on the air lift principle, where compressed air is fed from a quayside compressor to the helixors through a flexible plastic pipe anchored to the bed of the dock. As the compressed air is released into and rises through the tube, it drags water

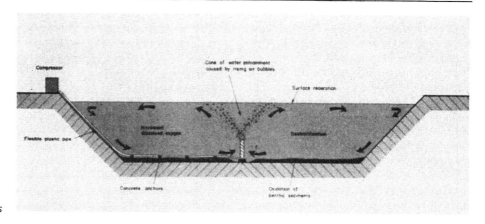

Figure 8.3 Helixors

into the tube and sets up a slow circulation of water within the dock basin. It is this circulation of water which moves water from the bottom of the dock to the surface, and provides aeration of the water and prevents stagnation (see Figure 8.3).

The system was initially operated twenty-four hours a day to bring about rapid improvement in the oxygen level. It was envisaged that a period of at least two years would be required to achieve an oxidized layer at the sediment surface. Once this had been established, operation could be reduced to intermittent periods in the summer, when it would be needed to prevent water stratification, maintain oxygen supply to the benthos, and control algal growth.

Changes in water quality

Separation of water inside the quays from the water in the adjacent ship canal resulted in a substantial reduction in pollutants, including reduced levels of suspended solids, ammonia and nutrients. The first indicator was that bacterial pollution reduced dramatically. This was primarily a result of separation from the polluted water in the ship canal. With the installation of the mixing system, the water column was de-stratified and an oxygen rich environment was achieved throughout the water column.

Development of the thriving aquatic ecosystem, the third element of the management strategy, was then begun. Some of the first organisms to benefit from the improvements were the algae, which are the starting point for many aquatic food chains. The zooplankton population increased rapidly, forming the next link in the food chain. An increase in the number and diversity of animal species soon followed, which is one of the best indicators of progress toward ecological stability. There has been a shift from a limited number of pollutant-tolerant organisms, such as worms and leeches, to an abundance of clean-water species, including water boatmen, snails and freshwater shrimps.

A key factor in this success has been the habitat diversification via the introduction of water plants and artificial reefs. The physical structure of the canal prevents the natural colonization of water plants, but this has been overcome by providing platforms and anchored baskets for water lilies.

The next element of regeneration of the water was the introduction of a fish population. The last salmon to run in the River Irwell, which becomes the Manchester Ship Canal, was caught in the 1850s. Since then the canal has remained virtually devoid of fish. Fish trials were carried out and fish stocking was undertaken between 1988 and 1989 and over 12,000 coarse fish were introduced to the quay. Subsequently a high growth rate has

Figure 8.4 View of Salford Quays today

been recorded in most species, with roach, perch and carp adapting well. In 1993 trout was stocked in the basins.

The water area is also frequented by a number of birds. The harbor area has been visited by cormorants and great-crested grebes. Only one water-quality problem still requires attention – that of blue-green algae blooms fueled by the nutrient-rich conditions in the basins. An environmentally friendly means of algae control is being pioneered which will result in long-term biological stability.

Salford Quays now

The strategy of the development plan to improve water quality and create attractive, visual landscape areas has resulted in many other activities now being carried out on the quays. The clean water is a suitable host to a wide range of water-based recreational activities. There is a water sport center which has activities such as sailing, wind surfing, canoeing and even skin diving. The university boat race between Manchester and Salford is now hosted by Salford Quay and pleasure craft are a regular sight (see Figure 8.4).

Overall the success of the strategy has been remarkable. The rapidly developing aquatic ecosystem has been central to this achievement, with the quays now reporting one of the fastest-growing fish populations in the UK. This was recently recognized by the Institute of Fishery Management and Angling Foundation, which awarded Salford Quays the most prestigious fisheries award: The Good Management Award.

The future for Salford Quays

Regeneration of the area continues with a landmark development led by the City Council to commission the internationally renowned architects Sir James Stirling (now deceased) and his partner Michael Wilford to prepare a master plan for one of the piers. This includes the provision of a performing arts center and a gallery to house the City Council's Lowry Arts Collection (Note: The Lowry Centre was opened in 2000). With lottery funding this proposal is being taken forward and will be named the Lowry Centre. Situated in the heart of the North of England's regional center, this site has planning permission and is within one hour's travelling time for 7 million people.

The footbridge across the Manchester Ship Canal will link the New Lowry Centre with Trafford Park. It is also now confirmed that the Metrolink system will be extended to the quays and it is hoped that the extension will be connected to the Lowry Centre. A plaza, new access roads and plentiful secure parking will make the Centre easily accessible.

It is forecast that the construction of the Lowry Centre will stimulate the development of the remainder of this area and have a catalytic effect on the development of the adjacent docks and Trafford Park.

Conclusions

Salford Quays is unique in terms of its scale and quality. The regeneration of a run-down, derelict dockland area has been achieved and it has become a trigger for wider regeneration. This has only been possible because:

- the local authority took responsibility for overseeing the reclamation, enhancement and development of the area;
- a development plan and program was prepared that not only had vision and clarity, but was achievable;
- quality in the highest standards of design and building were recognized as central to the overall development of the site;
- there was a technical team, including all of the relevant players, working closely with the client from the beginning of the development;
- the client and the team had long-term faith that the overall objectives could be achieved despite short-term problems;
- There was the willingness to recognize that a major project would take 15–20 years to complete and mature, and that this timescale would outlive most government funding programs.

Stockley Park Project

Introduction

The redevelopment of Stockley Park represents the successful transformation of a large area of land which had been despoiled and made derelict by gravel workings and subsequent waste disposal (principally municipal refuse disposal). The redevelopment has resulted in environmental improvement, the control and management of residual contami-

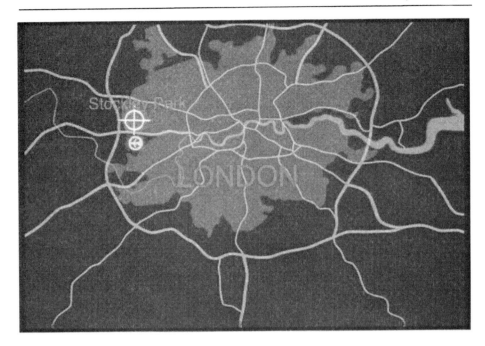

Figure 8.5 Site location plan

nation hazards, and the provision of beneficial uses of the land, both for the local community and for new businesses. Particular features of the project are:

- the partnership between the Local Authority and the Developer Consortium to realize a financially viable and practical scheme, taking into account the aspirations of local community groups;
- the engineering and reuse of on-site resources to minimize the transport impact of the reclamation works;
- the management of groundwater flows to minimize leachate generation;
- a substantial improvement in water quality and diversity of aquatic life in an adjacent water course;
- improved transport links and recreational access across the area.

Site description

Location
The site encompasses some 350 acres of land on the western fringe of Greater London, 25 miles from the center of the city (see Figure 8.5). This is an area where London's suburbs end and the green belt begins. The M4 motorway, which connects London to South Wales, and the M25 orbital motorway around London, are both within a mile of the boundary of the site. It is also very close (just 2 miles) north of Heathrow Airport and is bounded to the south by the Grand Union Canal.

Site history
From about 1850 onwards, the near surface gravels that occurred at the site had been worked in a piecemeal fashion to support building development in the area. This

Figure 8.6 Landfill operations at the site

continued for more than a hundred years (until the 1960s), by which time the majority of the gravel resource had been exploited. The filling of the gravel pits with solid waste commenced in 1916 and continued for almost seventy years (see Figure 8.6).

There was very little statutory control of waste tipping in the UK before the introduction of the Control of Pollution Act in the early 1970s. Consequently there were no reliable records of the nature of wastes on the site, which was known to have received illegal "fly tipping" from many sources. By the early 1980s all the refuse had been covered by a thin, poor-quality soil capping layer containing significant amounts of demolition rubble, which was ineffective in reducing surface water infiltration. The landscape presented a desultory picture of sparse vegetation, and consequently the site offered no amenity value and was the source of considerable environmental pollution as shown in Figure 8.7.

Figure 8.7 Landscape prior to redevelopment

Figure 8.8 North–south vertical section prior to reclamation

Ground conditions pre-reclamation

A typical vertical section north–south through the site as it was prior to reclamation is presented as Figure 8.8. At depth the site is underlain by a stiff plastic clay (the London Clay), which provides hydraulic separation from deeper deposits. Overlying the London Clay were the remnants of unworked natural gravels, in turn overlain by refuse varying in thickness up to a maximum of 12.5 m, and a thin soil capping layer. The site had a shallow overall slope from north to south, where it adjoined the canal. Natural groundwater levels in the area were 1–2 m below ground level and flowed from north to south. Since this large, essentially flat site was unlined and inadequately capped, large volumes of leachate were generated. The flow of this leachate from the site into the Grand Union Canal was causing serious water pollution. The main route for this pollution was the direct discharge of leachate from springs and pipes into the canal or connected water bodies. Some leachate probably also passed through the canal lining under the original hydraulic gradient. This pollution had been identified as "the Stockley effect" by Thames Water and British Waterways and was evidenced both by water quality testing and by periodic fish kills. Stopping this pollution was an important consideration of the Stockley Park Project. An additional problem that required management was the generation of soil gases, notably methane, from the decomposition of the more recent refuse.

Joint vision for site reclamation

The Local Authority (the London Borough of Hillingdon) owned the whole of the site, but did not have funds available to deal with the scale of environmental problems that existed. However, a private developer consortium recognized that the site had combined assets of size and location potentially to counterbalance the environmental constraints and liabilities. Following feasibility surveys and negotiation, the Local Authority and Developer Consortium reached agreement on the integrated remediation and redevelopment of the site. Under this agreement the Developer was obliged to reclaim the whole of the site and to

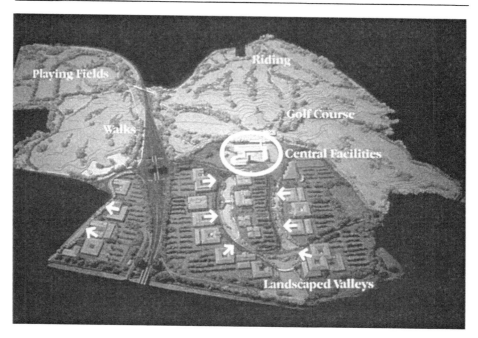

Figure 8.9 Land-use plan of reclaimed site

provide a Country Park for public use on the majority of the land. In return the Developer was given permission to construct a low-density technology Business Park on the southern side of the site.

The principal objectives of the reclamation strategy were to achieve substantial improvement in the environmental conditions of the site as a whole, including the pollution in the Canal, to bring a majority of the site back into public use without any associated health risks, and to provide totally uncontaminated sites for the buildings in the Business Park areas.

Following consultation with the local community, the Local Authority determined that the Country Park, which would occupy some 200 acres in the central and northern parts of the site, should comprise a championship standard golf course, playing fields, bridleways and other recreational facilities. The Business Park, extending over 150 acres, was to be constructed in three phases containing high-specification office buildings placed in a generous landscape setting. A land-use plan of the reclaimed site is presented as Figure 8.9.

Reclamation works

Development of remediation strategy

The Developer appointed Ove Arup and Partners to undertake the design of, and subsequently to supervise, the remediation works. Other principal members of the Developer's consultant team were the Ede Griffiths Partnership, landscape architects, and Grontmij R.V/LRDC International Ltd, reclamation consultants.

Extensive site investigations were initially carried out to determine the geotechnical and chemical ground conditions across the whole area of the site. This information proved crucial to the successful development of the reclamation strategy. In particular the investigations provided data on the thickness and gassing potential of the wastes, the groundwater regime, and revealed that lenses of unworked natural gravels remained in place below the waste in certain areas.

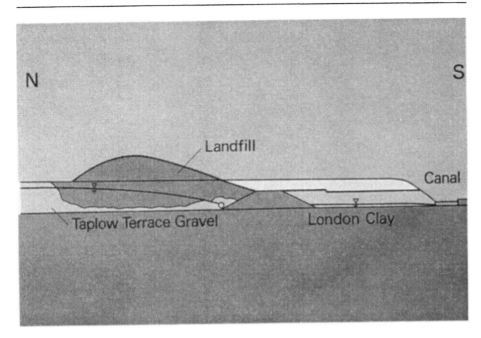

Figure 8.10 North–south vertical section after reclamation

An essential element of the design was to provide an uncontaminated site for the Business Park, and therefore required the mass removal of the wastes from the southern areas of the site. Export of this waste from the site and import of clean material was not possible, both because of the excessive cost of such an operation and also because it had been agreed that the construction traffic impact of the redevelopment should be kept to a minimum. A strategy based on optimum reuse of on-site materials and material swap from north to south was therefore developed.

Scope of the remediation works
A series of interrelated civil engineering works were carried out to provide suitable ground and surface conditions, to reduce the volume of leachate produced, and to control the sources of pollution which necessarily remained on-site. These works comprised:

- bulk earthworks and wastes relocation;
- installation of groundwater and leachate management systems;
- creation and placement of appropriate landscaped capping; and
- landfill gas management.

Earthworks and wastes relocation
A major earthworks contract was undertaken to utilize the available on-site materials to create suitable ground conditions and landforms. This involved:

- Stripping the capping materials and setting them aside for reuse (see pp. 96–97).
- Excavating all the refuse in the Phase 1 Business Park area (on the south side of the site) and placing it in the north (in the Country Park area).
- Scavenging natural gravels from beneath waste where significant residual thicknesses existed and reuse of the material in the creation of gravel pads in the south of the site to support the Business Park buildings.
- Creation of a clay borrow area in the north of the site and transport of clay from this

Figure 8.11 Aerial view during construction works

excavation to form bunds around the gravel pads and fill in landscaped areas. (The excavation of this clay also provided additional void space in the north for redeposit of the waste excavated from the south of the site.)

A typical vertical section through the site after waste relocation and construction of the gravel pads is presented as Figure 8.10. In total more than 4 million cubic meters of material have been relocated on the site. During the Phase 1 works the project is believed to have been the largest earthworks contract ongoing at the time in Europe (see Figure 8.11).

Groundwater and leachate management

As described earlier, the shallow groundwater had a natural hydraulic gradient across the site from north to south. As the groundwater passed southwards through the refuse it became heavily contaminated and large quantities of leachate resulted, which flowed into the Grand Union Canal.

To provide a permanent solution to this pollution problem the following two measures were installed:

- an in-ground vertical cement/bentonite cut-off wall around the north side of the Country Park;
- a leachate collection drain on the southern (down-gradient) side of the wastes.

In addition, heavy compaction of the wastes and profiling of the surface was specified to increase surface water run-off and reduce infiltration.

The cut-off wall is 600 mm in thickness and is keyed into the underlying London Clay. Consequently clean groundwater is now forced to flow around the eastern and western sides of the Stockley site, or be carried by a pipe which flows north–south from the center of the wall, through the Business Park, and to a discharge point into the canal on the south side of the site (see Figure 8.12).

The leachate collection drain comprises a perforated pipe in a gravel trench extending along the outer face of the clay bund walls around the Business Park with spurs to the south and four intermediate pumping stations (see Figure 8.13). The pipework and pumps

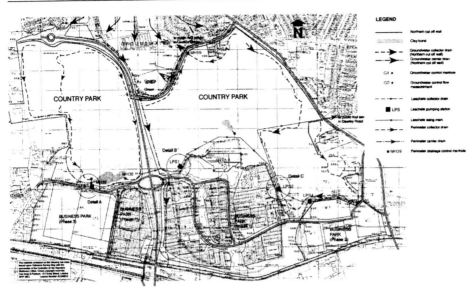

Figure 8.12 Plan of groundwater and leachate management systems

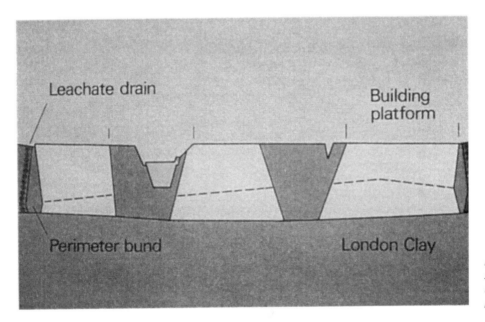

Figure 8.13 Vertical section indicating leachate drain, building platforms and bunds

control the groundwater level at a slightly lower elevation than the level of water in the Grand Union Canal. A hydraulic gradient is thus provided away from the canal to the leachate drainage system, reversing the previous direction of flow.

The leachate collector drains also receive stormwater run-off from the land drainage system of the Country Park. Excess leachate is transferred, via a rising main, to a local sewage treatment plant. The demand on the pumps is such that they have not been required to operate continuously.

The diversion of groundwater through-flow, together with the accumulation and compaction of the waste into a smaller plan area, the provision of a thicker and more profiled and vegetated cap in the Country Park area, and an increased density of the waste, has very significantly reduced the volumes of leachate generated by the site. Also the threat to the quality of the water in the Grand Union Canal has been eliminated by the hydraulic gradient reversal, removal of excess leachate to a treatment works, and the physical barrier of the bunded and uncontaminated Business Park between the canal and the waste.

Landscape development in the Country Park

Objectives
The long-term objectives for the Stockley Country Park were to create:

- a public open space and landscape that contains semi-natural habitats where natural regeneration is encouraged;
- a range of recreational facilities within a country-like environment for the pleasure and relaxation of the community.

The major elements of the landscape design within the context of site reclamation were:

- landform and topsoil construction;
- planting and seeding.

Landform and topsoil construction
The new landform for the Country Park, including the golf course, has been created by the transfer of refuse from the Business Park area. A topographically interesting landform was constructed on what was previously a flat area.

The successful development of vegetation in the Country Park required management of landfill gases and the design of an appropriate medium in which trees, shrubs and grass could be grown. The nutrient, moisture and physical support demands of trees, shrubs and grass vary, but construction considerations dictated that different topsoil construction profiles for different vegetation types were impracticable. Therefore a uniform "maximum" topsoil construction depth capable of supporting trees, shrubs and grass was adopted over the whole area.

For the bulk filling operations the newer refuse was placed below the older refuse, in order to reduce the risk of damage to vegetation due to methane generation.

The subsoil and topsoil capping layer for the Country Park was manufactured using, as far as possible, materials already on-site. This avoided the necessity of importing large and expensive quantities of topsoil.

The available on-site materials were essentially the refuse and the capping which had been scraped off the refuse areas. In addition, sludge cake with low levels of heavy metals was available from two local sewage treatment works.

In essence, the growing medium consists of two layers. The lowest layer, 750 mm thick, is composed of Type B former capping material mixed with aged refuse. Above this, Type A former capping material, mixed with sludge cake, was placed to a thickness of 550 mm. The basic difference between Types A and B is that the former contains no more than 40 percent gravel and stones, as a result of stone picking, undertaken during construction, to improve the quality of the final product.

The water retention capacity of the soil profile has been estimated at 250 mm which, together with the average rainfall of 250 mm during the growing season, provides adequate water for plants during dry periods when the groundwater level is low.

Planting and seeding

The long-term objective of the tree and shrub planting was to create the blend of species and age structure of oak, ash and hornbeam woodland which occurs naturally on clay-based soils in lowland England.

At Stockley Park the achievement of a mature woodland is being compressed to about a hundred years by planting tree types from all the stages of natural colonization at the same time. This increases the chances of success over only planting a few species and gives the option for the creation of woodland variety by the removal of different species at different times in different locations.

In view of the particular difficulties in ensuring healthy growth on a rubbish tip, the following techniques were used to increase the likelihood of success:

- Fallow-seeding the whole site with grass and clover to create a temporary cover. This was maintained for at least one year in order to bind the soil and prevent erosion, improve soil structure and create organic matter. This cover was also an "indicator" to show up areas of methane generation.
- Introduction of 1.6 million worms to improve soil structure and further activate the soil-forming processes.
- Creation of a "bare earth" soil cover regime in plantation areas; i.e., fallow grass and other herbaceous plants are prevented from growing in the plantation for about five years. This prevents competition with young trees and shrubs for soil nutrients and moisture.
- Conversion of the temporary fallow grass and clover into permanent grass and wild-flower areas by ploughing-in to incorporate organic matter into the soil, and reseeding with various grass and wildflower mixes to create permanent cover.
- Planting of native trees and shrubs which are likely to survive in landfill conditions. These are species which would develop naturally or colonize the site over a long period if the site were left in an unplanted state.

Over 140,000 trees and shrubs were planted as 2-year-old transplants, 300 mm to 450 mm high. These transplants more easily adapt to difficult conditions and therefore grow more quickly and healthily. To help planting establishment and maintenance operations, trees and shrubs were planted in a regular grid pattern; this will disappear in time.

Over 4,000 semi-mature and mature trees have been planted in key locations and contribute to the overall Park structure. This includes avenue planting which emphasizes major footpaths that cross the Park, linking different areas or focusing on special features such as the "Circle" in the District Park or the Golf clubhouse. These larger trees were planted in individually excavated pits.

Grassland covers approximately 80 percent of the Park area and different maintenance regimes reflect and direct the intensity of use, providing visual variety and additional habitats, particularly in the transition zone from grassland to woodland edge. At one end of the spectrum there are manicured trees and greens; and at the other, rough grass and wildflower areas. The wildflower areas provide ecological diversity and visual interest within the Park. It is anticipated that wildflowers will spread into other areas of the Park and maintenance can be varied accordingly. The habitat enrichment is part of the overall strategy for vegetation development in the Park and provides educational interest.

Management of landfill gases

The process of decomposition of rubbish generates gases amongst which are methane, carbon dioxide and hydrogen sulphide. Whilst these gases are harmless when they escape to and dilute in the open air, they can become dangerous when they collect in a confined space. Methane in particular forms a potentially explosive mixture with air when the relative quantities of the two fall within a certain range. Furthermore, the presence of methane and other gases within the root zone of vegetation damages plant life by creating the adverse conditions of oxygen starvation and toxicity.

At Stockley the control of gases is achieved by a combination of methane barriers, positive extraction and passive venting. In the Business Park the gas problem has been resolved by the removal of source materials (the rubbish) and the construction of very-low-permeability perimeter clay bunds as a barrier to lateral ingress, with a vent trench on their outer face.

In the Country Park area the gases are positively extracted, as required, by a series of wells, and also passively vented on the southern side by the gravel backfilled leachate drain trench.

Positive extraction system
There are fifty-five gas extraction wells located within the Country Park which penetrate the ground up to 16 m in depth. They are connected by polyethylene pipelines to a central pumping station. The methane is burnt off in a flare system, which is concealed within a screen. The flare system is capable of venting up to 1,100 m³ of gas per hour, which is well in excess of the recorded demand. The potential use of the gas as an energy source was examined but found not to be viable.

Passive venting
Stone-filled trenches, which at depth contain the leachate collection drain, provide a pathway for any landfill gases which are not actively extracted by the wells to disperse safely to atmosphere along the southern side of the Country Park where it abuts the Business Park (see Figure 8.14). The line of these trenches is well away from any buildings (which are all inside the clay bunded area), and the surface of the trenches is maintained free of vegetation. Monitoring has shown that landfill gas concentrations in the upper layer of the trenches are typically very low.

Figure 8.14 View in the Country Park (a)

Success of the reclamation

The success of the reclamation works can be measured in terms of the environmental improvements that have been achieved, the community benefits, and the thriving business technology park that has been created.

Environmental improvements

Many different bodies were involved in the design and approval of the remediation works. In addition, monitoring of various pollution indicators occurs on a regular basis to confirm the success of the reclamation strategy.

Forty permanent boreholes have been established in the Business Park, Country Park, and outside their perimeters, where long-term monitoring of the quality of water, together with presence of gases, is being undertaken.

Monitoring of water quality in the Grand Union Canal has provided clear evidence of the success of the project. As previously noted, prior to the construction of the Stockley Park Project, the leachate flowing from the site was causing serious pollution in the Grand Union Canal. This pollution had been identified as "the Stockley effect" by Thames Water and British Waterways and it was an important consideration of the Stockley Park Project to stop this pollution.

Two good markers of this "Stockley effect" were the ammonia concentration in the canal water and its oxygen demand. Prior to reclamation ammonia concentrations increased from 1 mg/l to 5 mg/l as the canal flowed past the site, and there were regular fish kills. Following completion of the reclamation works, the concentration of ammonia was found to have declined from 1 mg/l to less than 0.5 mg/l as the water flowed past Stockley Park. The chemical oxygen demand and biological oxygen demand have also shown a dramatic reduction.

In the Country Park area the success of woodland planting and grass establishment are the most visible and tangible measures of success of the reclamation process (see Figures 8.14 and 8.15). It indicates that the methane issue has been resolved and that a successful drainage and growing regime has been established. Fallow grass and clover establishment

Figure 8.15 View in the Country Park (b)

*Figure 8.16 View within
the Business Park*

was completely successful. Highly localized and "patchy" areas of methane generation occurred, particularly in areas of relocated "young" landfill. These initially covered no more than 5 percent of the total area, and this fell to less than 1 percent when the gas extraction system became operational.

Tree and shrub planting has been exceptionally successful. In four years, trees have grown from 300–450 mm to 2–3 meters high. On landfill sites failure rates are commonly up to 30 percent and sometimes more. Of over 140,0000 trees and shrubs planted at Stockley Park, the failure rate was only 7 percent. This is quite exceptional and is attributable to the success of the integrated land engineering, reclamation, topsoil construction, landform, drainage and planting regime, and to high standards of "land construction" by strict adherence to a high specification and a staged, paced implementation strategy.

Community and business

Stockley Park is now a thriving Technology Business and Country Park (see Figures 8.16 and 8.17). The Phase 1 buildings on the site total 1.4 million square feet gross floor area. Thirty-four companies currently occupy this commercial space and 6,650 people are employed. Phases 2 and 3 of the Business Park are planned to add another 0.9 million square feet.

The recreational facilities on the Country Park are widely used by the local community, as well as by visitors from further afield. The new north–south road through the center of the site also provides an important transport link.

Conclusions to both case studies

The two case studies are clear examples of sites with serious constraints and liabilities being "manufactured" to new beneficial uses. The scale and location of both projects was such that it took considerable vision to commit resources to the sites, and a lot of faith and patience to see the projects reach fruition.

The success of the projects have been realized in many ways, including control of pollu-

Figure 8.17 Aerial view of the Business Park

tion; long-term employment prosperity, provision of public recreational facilities and new homes; biodiversity of natural fauna; and sustainable use of the site's natural resources. Several key factors combined to make these achievements possible. Not least amongst these was the commitment of the client and regulatory authorities, and the multidiscipli-nary professional teams.

Acknowledgements

The authors acknowledge the assistance and support of Salford City Council, Stockley Park Consortium and Hillingdon Borough Council in producing this essay.

Response

Integration through an interdisciplinary approach?

The two case studies, Salford Quays and Stockley Park, presented by Lorna Walker and Richard Owen, respectively, offer compelling examples of how the multidisciplinary approaches used in both projects were critical to the regeneration of these large-scale, severely degraded sites. It was determined in both projects that a mixed-use solution with recreational amenities would provide economic viability. A water sports center and an arts center were proposed in Salford Quays, and an extensive recreational park and business park were embraced by the community and implemented at Stockley Park. The scale of the projects, the severity of the contamination and the complexity of the mixed-use programs at both sites required a diverse team of designers, planners, economists and scientists to bring their shared expertise to the table and integrate the remediation of the land and water with the economic and civic goals of the projects.

The two sites are contextually very different. Salford Quays is dominated by water requiring extensive hydrological and aquatic remediation. Stockley Park, a former inland gravel pit, later developed as a solid waste dump, necessitated extensive subsurface groundwater and leachate mediation, regrading and soil treatment, and gas control and extraction. They both raise some common site specific issues: notably their site remediation and programmatic developments are closely tied to market rate economics. A more expanded look at the economics is found in an overview of urban development related issues in the essay in Chapter 6, "Engineering Urban Brownfield Development: Examples from Pittsburgh," by McNeil and Lange. In the latter study the funding partnership between the public and private sectors, and the reliance upon interdisciplinary design teams, is elaborated upon. In Salford Quays, improved water quality was fundamental to any future revitalization efforts, and several innovative technologies were created to enhance the water quality and assure that all economic and building development could follow.

As is common in most remediation projects the pivotal question to resolve is how multiple problems can be re-seen and addressed to become complementary assets. The methods used in both Salford Quays and Stockley Park reflect this multiple solution approach required to meet a complexity of issues. Instead of using a single technology, the design team in the Salford Quay project incorporated multiple methods to begin to improve the water quality using technologies to simulate natural features. Strategies include artificial reefs, structures to support aquatic plants and the introduction of oxygen and aeration; these efforts grounded in both environmental engineering and the science of ecology. By introducing the aquatic life and utilizing the natural food chain to increase the clean-water species, the team designed a long-term management plan aimed at continued improvements in water quality. Innovative solutions were developed by a close collaboration between designers, engineers and scientists. Recreation added a key and complementary component. The remediation of Salford Quays with costly water quality improvements was economically acceptable in part because recreational opportunities were created. In Stockley Park, multiple benefits were also critical to acceptance of the project by the local community. These include a golf course, pedestrian and bicycle trails, ballfields and small playground facilities

A full and accurate understanding of the site contamination early in the design process was important in both projects. The multiple methods that were used in Stockley Park responded not only to the flow patterns of water migrating through and off the site, but also to the differing soil densities and specific locations of contamination. An effective and appropriate technological response was tied to information gathered about water contamination, sources, flows, distributions and concentrations. At Salford Quays it was determined from water quality data that run off had to be separated from the main canal to restrict ongoing contamination. The solutions are ultimately only as effective as the information available at the time.

Because the Salford Quays had an existing infrastructure, specifically its remnant dock walls, its development responded, in form, by a lineal spatial organization and programmed uses respecting the historic precedence of the site. However, instead of restoring these forms and activities to replicate past uses, which were largely economically obsolete, the designers and planners adapted the existing structures to create cultural and economically appropriate activities. Many are water-based activities, offering a wide range of recreational uses. The docks are integrated into the site as promenades for strolling. The inclusion of the arts center will act as a civic anchor for future development. While the designers did not recycle the historic structures as Peter Latz has done in Chapter 11, in part because so little was intact, Salford Quays does respect the spatial qualities of the historic development and in so doing a referential memory is preserved. The challenge in Salford Quays was to re-establish a healthy ecology respectful of the spatial qualities established when the port land was first developed. The strategy used in the Stockley Park project is dramatically different. While a strong attempt was made to reuse and redistribute existing materials largely for economic reasons, this site has undergone a complete transformation with topographic changes and the introduction of massive numbers of trees, shrubs and grasses. Through the relocation of the site materials, references to prior uses were removed. In this project the designers responded not to existing forms but instead to existing conditions, selecting for example what plant species would be best matched to soil conditions and varying degrees of contamination, and deciding how to rechannel the water flows across and through the site to optimize its biological health.

It is interesting to see that while the sites had long been available, timing was critical to the project development and each project was possible only when the right conditions, including community support, financial investment and appropriate remediation technologies, came together. In Stockley Park the access to the airport and motorway were critical to the success of the recreational and business park. When reviewing the plans for Stockley Park, the community perceived the addition of 200 acres of recreational land as an asset, and the availability of land and development potential envisioned by the private developer were critical conditions that had to coexist at the same time. The issues of timing and development at Salford Quays differ from those of Stockley Park. While both were dependent on private/public financial support, it was critical that all the participants understood that this project would not result in an immediate transformation of the site (as was the case at Stockley Park), but instead that the timetable for full recovery both ecologically and economically would take place over a longer time-frame. It would be a long, incremental process, a process different from most traditional development projects and one that requires a sequential process of development and investment. Because of the longer time-frame a comprehensive plan was critical to guide future development and insure that additions were integrated into the context and supported the vision that would steer the revitalization of the area.

Both these projects must be seen not simply as site-specific remediation projects, although they clearly addressed the contamination issues inherent to the respective sites. Each offers greater gains at a regional level, including continuing development interest in both these areas. The open space as a drawing card is significant in both projects, but most noticeably in Stockley Park where the reclaimed open space developed as a recreational opportunity had a regional impact, one that should be integrated into current and future regional open space plans. It may be that in the future contaminated sites should be mapped as potential open spaces when planners and designers study regional open space master plans in these types of regions. This is not unlike the thought process used by the team engineers as they mapped the flows off-site and had to determine the impacts of the contaminants well beyond the site boundaries and the impact to both surface and underground water bodies well beyond the site. Thus in these projects the scale of the design must also include regional planning as an important component.

Chapter 9
Industrial evolution:
prevention of remediation through design

Jean Rogers

Introduction

In his book *The Environmental Economic Revolution: How Business Will Thrive and the Earth Survive in the Years to Come*,[1] Michael Silverstein points out that some 2 million Americans are already employed in the field of environmental protection.

Annual expenditure on environmental clean-up in this country alone is $130 billion, and there is another $50 billion international trade in remediation. In this decade, $1.2 to $1.5 trillion will be spent domestically on clean-up. The travesty is that the burgeoning of this domestic and international market could have been entirely prevented. Companies are beginning to learn that being energy-efficient, materials-conscious, handling their manu-

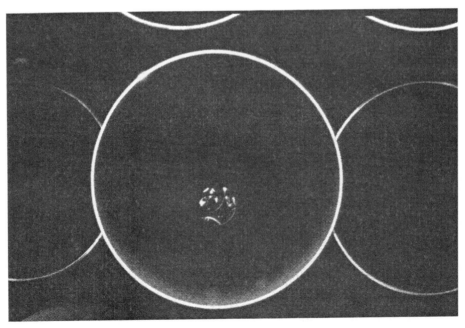

Figure 9.1 Napa Pipe, 1991

facturing processes responsibly, and creating durable, high-quality goods will not only avoid remediation costs but also drive competitive advantage in the marketplace. In the 1970s, industrial concerns were focused on end-of-pipe treatment, and coping with compliance of the seemingly interminable proliferation of environmental regulations (RCRA, CERCLA, CAA, HSWA, etc.). In the 1980s, waste minimization took off, and the three Rs mantra was heard emanating from all but the most recalcitrant corporate environmental health and safety departments—Reduce, Reuse, Recycle. In the 1990s, Design for the Environment (DfE)[2] caught on as the latest management tool enabling line managers to focus on what they always should have been focused on– operations, while still doing the right thing for the environment. It is clear from the improved profile flaunted by products, processes and facilities that shifting the focus from crisis management to proactive strategic management of environmental issues has been a boon for business, design, and the environment alike. These successes have been achieved by a handful of progressive companies, with the vision to look beyond running the numbers, to running the company in a sustainable manner into the next century.

Through highlighting a number of industrial projects this essay will advance the notion that protection of the environment and preservation of the landscape is best achieved through comprehensive, intentional design of industrial facilities, and not through remediation. The purpose of this essay is not to establish a dialectic between design and remediation, because in many cases, when environmental degradation has already occurred, remediation is the only viable solution. However, this complex and costly process can be avoided in many cases through environmentally sensitive design and operation of industrial facilities. This is a lesson that in many cases has been learned the hard way, in the face of stiff fines and penalties, or through the creation of a Superfund site, the legacy of which only a Potentially Responsible Party (PRP) can truly appreciate. Napa Pipe Corporation, a subsidiary of Oregon Steel, (Figure 9.1) struggled to remain competitive in the marketplace while addressing historic environmental concerns. They ultimately found a way to rectify these competing forces to their advantage, and to mitigate the effects of industrialization on the environment.

Pipe dreams

Napa Pipe is located in the lovely and scenic Napa Valley in northern California, on the Napa River, at the beginning of the well traveled wine tasting route from San Francisco to the abundant regional vineyards (Figures 9.2 and 9.3).

Figure 9.2 Napa River, 1995

Figure 9.3 Napa Pipe Plant, 1997

Wastes from manufacturing operations were disposed of on-site for about fifty years.[3] As part of the traditional steel pipe manufacturing process, welding flux laced with toxic heavy metals is generated: cadmium, nickel, mercury, lead, and copper. The flux had no economic value, was structurally stable, and would have made an excellent fill material had the metals concentrations not been high enough to constitute hazardous waste. The groundwater, flora and fauna in the wetlands were severely impacted by the heavy metals leaching from the adjacent disposal area

The goals of the remediation project were to renew the wetlands and prevent future harm, as well as to create a viable pipe storage area for the finished product. Removal of the wastes was one of the alternatives considered as part of the extensive feasibility study and risk assessment conducted by the client. However, moving 50,000 cubic yards of material elsewhere would have been ecologically unsound, and at $500 per cubic yard it would have cost $25 million and meant filing for Chapter 11 bankrupcy proceedings for Napa Pipe. As the single largest taxpayer in the county, it was in no one's best interest to put them out of business. These programmatic constraints led to the design and construction of a landfill that would contain the wastes, protect the wetlands *and* could be constructed in place (Figures 9.4–9.8).

Integral to the design was a vegetated gabion wall, which provided slope stabilization and eased the visual and physical transition from the submerged wetland to the paved pipe storage plateau. A metals precipitation and treatment system for the captured groundwater was also incorporated.[4] Contaminated soil and organic material was dredged from the adjacent wetlands and treated on-site over a three-year period (Figure 9.9). A massive PVC-lined bed was constructed to hold the material – 10,000 cubic yards of sediment laced with heavy metals and semi-volatile organics (Figure 9.10).

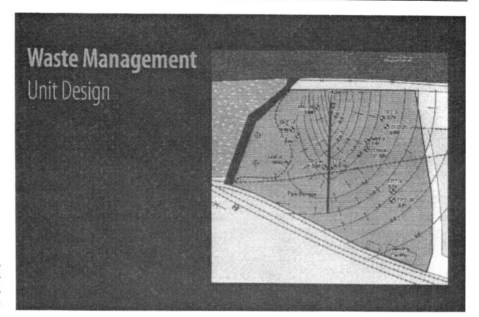

Figure 9.4 Waste management unit design, Napa Pipe Corporation, 1992

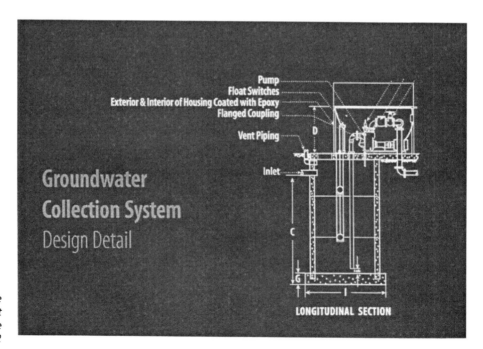

Figure 9.5 Waste management unit design, Napa Pipe Corporation, 1992

Figure 9.6 Construction of waste management unit, 1996

Figure 9.7 Construction of waste management unit, 1996

Figure 9.8 Construction of waste management unit, 1996

Figure 9.9 Dredging operations, Napa Pipe, 1990

Figure 9.10 Bio-remediation staging area, Napa Pipe, 1990

Detoxification of the material was initiated through successive biological and chemical processes – first, aerobic oxidation of the organics (tilling to aerate the soil and enhance natural microbial counts, as well as to mix in the nutrients nitrogen and phosphorus). Metals were subsequently stabilized and fixed through the addition of a lime-based mixture.[5] The soil undergoing transformation was shaped into long narrow furrows for the duration of treatment – evocative of a meticulously sculpted local vineyard, perhaps. The treated soil was recycled for use in landscape applications on-site.

The wetlands before and after remediation are shown in Figures 9.11 and 9.12, the result of a capital expenditure (including engineering, design and construction) amounting to $5.2 million as well as $250k/year for maintenance and monitoring. Monitoring is required quarterly by the state in perpetuity in lieu of allowing wastes to remain in place.[6]

Figure 9.11 Wetlands before remediation, Napa Pipe, 1990

Figure 9.12 Thriving wetlands, Napa Pipe, 1992

Through applied technologies, the wetlands at Napa Pipe are protected. The permanent monitoring well network (Figures 9.13 and 9.14) enables the groundwater to be monitored for aberrations in metals levels resulting from a breach of the containment system. The hazardous wastes will remain in place forever or until the "Big One" hits California. The technology is omnipresent on-site, an integral part of the infrastructure: an underground network pulsating contaminated water through an elaborate system of subsurface conduits. The technological system is insured against changes in plant ownership, new management or sale of the property through deed and land use restrictions, amounting to a perpetual liability for the parent company Oregon Steel. The permanent presence of this elaborate and costly technological apparatus raises some important questions about the

Figure 9.13 Monitoring well, 1992

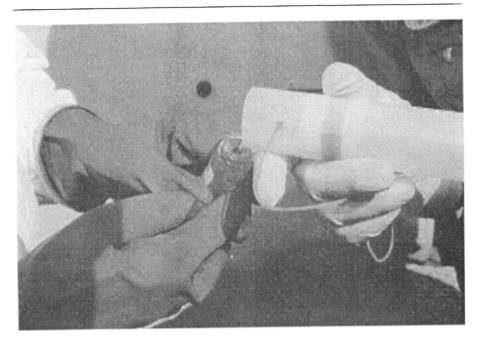

*Figure 9.14 Ground-
water sampling, 1992*

use of technology and the implications of the remediation process. What if these remediation funds had been invested in new manufacturing technologies or state-of-the-art
manufacturing facilities instead? Or spent on R&D, or increased wages or employee
bonuses, or education and safety, or art on the walls in the cafeteria, or any number of
other things besides manufacturing their site?

Napa Pipe did repent and change their ways. They have designed and installed a wastewater treatment plant, for 100 percent recycling of process water – no small feat, because
for every ton of steel that is produced 37 tons of water are consumed. They have also
switched to a non-toxic welding process, hired a director of environmental affairs, and put
an environmental management system in place, which includes policy and performance
measures, awareness and education, a strategic plan for competitiveness though environmental, and systematic assessment of the impacts of products and processes. Therein lies
the real success story. If nothing else, remediation, with its poor company press and time-
consuming process, inspires companies to make sure they don't make a habit of manufacturing sites.

Another response

As a counter to the "fix it when it's broke" approach most often taken by industry up
until the 1990s, this section will highlight projects that were executed correctly from the
beginning: manufacturing facilities and plants designed in such a way that it would be
unthinkable and even unfeasible to receive an environmental permit violation let alone
degenerate into a Superfund site. These projects demonstrate how design is a fundamental means by which the environmental profile of our products, processes and buildings can
be improved. Many designers, and companies worldwide have already made great strides
in lessening the environmental impact of their products, facilities, and manufacturing
operations through intelligent design, often with improved public relations, shareholder
value and reduced costs over the life cycle.

Born at the Bauhaus

Early in the century factories were beginning to be designed to integrate function with form. The Bauhaus modernist philosophy was embodied in Walter Gropius's 1911 design for the Fagus Shoe Factory in Germany, completed when he was just 28. The plant still manufactures shoes today, 90 years later – one definition of sustainable architecture. The plant is a model of industrial design and integrated manufacturing operations, and typifies the European philosophy – factories are built to last,[7] and the surrounding environment and community reap the benefits of the committed relationship. The German propensity for efficient operations and effective design continues today.

Enterprise as ecological system

Wilkhahn is a progressive and highly acclaimed furniture maker also based in Germany. The firm made its reputation through their commitment to cutting-edge design and the environment, from the factory floor to the finished product. "Buildings are answerable to the future," states Fritz Hahn, chairman of the board, who believes in enterprise as ecological system.[8] He called for economic, ecological, esthetic, and humanitarian standards in design of their factory. Architect Frei Otto used ecologically sound materials (such as the simple wood frame roofs) and incorporated such amenities as outside views and under-floor heating for seamstresses. Their eco-hotline enables specialists to address waste minimization, pesticide use in landscape and design, and eco-product development issues. This fosters a corporate consciousness and culture of respect for the environment which would render it unconscionable to pollute.

Miller meets McDonough

Herman Miller's Furniture Manufacturing facility in Zeeland, Michigan, completed in 1995, integrates landscape, attention to physical health and comfort through elimination of toxic materials, ventilation and daylighting. It exemplifies Herman Miller's attention to environmentally conscious design, also practiced in their sleek, durable, and recyclable office furniture lines. Architect William McDonough, who brought the concept of sustainability to the architecture profession, reconciled environmental concerns with an integrated vision of marketing, design, and production.[9] Waste recovery and recycling procedures were employed during construction. Environmentally sensitive building materials were used from eco-carpet, to formaldehyde-free particle board, to VOC-free paints and varnishes, to energy conscious glazing. In the landscape design, Pollock and Associates countered the environmental consequences of the building through a number of green strategies. Prairie grasses, man-made wetlands and swales filter stormwater run-off and treat graywater, and the meadow of mixed grasses provides a low-maintenance lawn.

One of the most significant aspects of the Miller facility is the commitment to quantifying the return on investment. The DOE and US Green Building council have undertaken a study to evaluate the effects of improved daylighting and indoor air quality on the incidence of sick days, manufacturing errors, and other aspects of productivity. It is the most sophisticated monitoring of this type to date in the US, and should provide ample justification for design of green facilities and processes beyond the feel-good factor.

Molto Benetton

Benetton has long preached and incorporated environmental principles into the design and manufacturing of its products, as well as into its state-of-the-art clothing factories. Benetton's commitment to the earth is a part of their corporate identity. Luciano Benetton himself has declared that the enterprise no longer has a merely economic function, but also a social, cultural, environmental, and political one.[10] Benetton was the first clothing manufacturer to employ information technology strategically. Benetton links point-of-sale data from the cash registers at its 700 retail stores in 120 countries with its production facilities in Treviso, Italy. Just-in-time manufacturing techniques enable production of the precise fuchsia sweater needed to replenish inventory wherever it is sold around the world, with minimum lead-time and zero waste. The facilities are designed to support this just-in-time philosophy, based on a continual flow of goods and information, as applied to the production system and waste management.

While color and cut remain seasonal, state-of-the-art green manufacturing processes are continually employed for cutting and dyeing at the Villorba complex. The factories and automated distribution center at Castrette were designed by Afra and Tobia Scarpa. Tadao Ando designed Fabrica, a recently completed research and design (R&D) facility testament to what can be achieved when profits need not be diverted to remediation. Benetton has made admirable progress, balancing the ephemerality of fashion with sustainability in design and incorporating environmental principles into all aspects of their operations while enhancing competitiveness along the way.

High tech/low impact

There are many forces at work shaping the design of industrial buildings around the world. Perhaps the biggest influence of all is the increasing need for flexibility and expandability.[11] Product life cycles are shorter and technological change is more rapid. The IGUS plastics manufacturing plant in Germany by Nicholas Grimshaw was designed from the start with the idea that the market may dictate that the facility be a plastics plant today, or a retail food store tomorrow.[12]

The building incorporates flexibility through design of the cladding and interior spaces. The open plan is a result of the unique suspension system and facilitates environmental control technologies. Waste management and recycling operations are incorporated into the layout. Technology in this case is used for prevention rather than cure. 3-Com in Santa Clara, California, is also militant about operational flexibility. The ability to flip space between office and manufacturing uses is key. Converting office floors to production lines requires clean, controlled production technologies and upgraded work areas with skylights, natural ventilation, and views.[13] Flow-through manufacturing made it easy to incorporate environmental controls and waste management technologies into the building, which in turn kept construction costs down. This is a fine example of a low-cost, agile manufacturing facility, which integrates concern for people and the environment, and provides a competitive edge.

This approach is uplifting given that 3-Com's plant is located in Silicon Valley, the largest Superfund site in the country. The groundwater in the region was contaminated in the 1980s with chlorinated solvents from irresponsible chip manufacturing. It will likely take more than a hundred years to clean up the aquifer in the region using pump and treat technologies, another high-tech legacy.

Designed for the environment

While these industrial facilities take many different forms and house different functions, a commitment to facilities in harmony with the environment is common to all. The examples cited have brought environmental technology inside the plant and wed them with production engineering, materials management, process layout, and product design. They have, in fact, designed their manufacturing sites well, to avoid becoming another remediation statistic. If there is one message, one lesson learned from the travails of the remediation industry, it should be that the focus of our efforts, our energy, and our creativity should be directed toward design of our facilities, processes, products and environments, and as an extreme last resort upon remediation. Environmental degradation should not be accepted

Figure 9.15 Drill rig, 1995

as a given. The uncontrolled release of toxic materials from our factories and landfills is not a new and exciting challenge to be incorporated into the design of the landscape. It is a far greater challenge to eliminate revealed and concealed impacts on the environment entirely. Designers are in the unique position of being able to eliminate environmental impacts from the start. This means considering the environmental implications at every stage of design, just as program, site, context, function, form, technology and material are considered. Involved are the hard decisions that designers face every day, such as facility siting and reuse options, building envelope design, energy considerations, materials selection, integration with operations, and, most importantly, communicating the design vision to management (Figures 9.15 and 9.16).

As we celebrate sites which have been manufactured in this compendium of essays, let it not be forgotten that we should be intelligently designing our manufacturing sites to begin with:

- Factories which are integrated with the location and site, and do not put at risk either the workers, community or the environment.

Figure 9.16 Trees, 1995

116

- Factories which conserve energy and resources, where water is appreciated as a precious resource not to be wasted in processes or landscape design.
- Facilities which account for the hazardous nature of materials handled and the toxicity of products produced, and expose piping above grade for ease of maintenance and prevention of unnoticed leaks into soil or groundwater.
- Plants which provide natural light and views and breed a sense of pride in the workplace.
- Facilities which understand human fallibility, and provide reassurance and insurance in their process design.
- Facilities in which technology is directed toward prevention rather than to the control of environmental impacts.

The examples cited all include one similar component from inception – respect for the environment. They were conceived as total ecological manufacturing sites – a symbol of an elusive industrial ideal in which work and well being, technology and conservation of the environment are finally linked. It is in this spirit that our sites are best manufactured. It will take nothing short of an industrial evolution.

Notes

1 Silverstein, Michael, *The Environmental Economic Revolution: How Business will Thrive and the Earth Survive in the Years to Come* (London: St. Martin's Press, 1993).
2 Rogers, Jean, "Design for Environment: Design Management Practices and Opportunities from Diagnostic to Design" (San Francisco, Calif.: Deloitte and Touche, 1997). This methodology assesses the environmental impacts of facilities and products during the planning and design phases. The framework is intended as a guide for design managers to incorporate environmental principles into design without sacrificing esthetics, function, or cost.
3 Montgomery Watson, *Site Investigation/Feasibility Study for Napa Pipe Corporation* (Walnut Creek, Calif.: Montgomery Watson, 1990).
4 Montgomery Watson, *Waste Management Unit Design – Napa Pipe Corporation* (Walnut Creek, Calif.: Montgomery Watson, 1992).
5 Montgomery Watson, *Treatability Study for Bioremediation – Napa Pipe Corporation* (Walnut Creek, Calif.: Montgomery Watson, 1991).
6 Montgomery Watson, *Quarterly Monitoring Report for Napa Pipe Corporation* (Walnut Creek, Calif.: Montgomery Watson, 1993).
7 Gordon, Barclay F., "The Fagus Factory: Contemporary Design Seventy Years Later." *Architectural Record* 169 (9) (New York: McGraw-Hill, 1981), pp. 114–17.
8 Brozen, Kenneth, "Wilkhahn's Direction as a Design Leader," *Interiors* 148 (6) (New York: Billboard Publications, 1989), pp. 120–3.
9 "Herman Miller," *Architectural Record* Vol. 12 (New York: McGraw-Hill, 1996), pp. 28–33.
10 Zardina, Mirko, "Benetton," *Lotus International* no. 85 (Milan, Italy: Elemond, S.p.A., 1995), pp. 101–21.
11 "Industrious Architecture," *World Architecture* no. 54 (London: Cheerman, Ltd, 1997), pp. 119–25.
12 Mark Bryden, "Factory with Flexibility Built In – The IGUS Plant," *The Architects' Journal* 197 (12) (London: emap Business Publishing, 1997), pp. 33–44.
13 "Manufacturing Collaboration – 3–Com," *Architectural Record* 12 (New York: McGraw-Hill, 1996), pp. 23–39.

Further reading

Barnett, Dianna and Browning, William (1995) *A Primer on Sustainable Building*, Snowmass, Colorado: Rocky Mountain Institute.

Crosbie, Michael J. (1995) *Green Architecture*, Rockport, Mass.: Rockport Publishers.

Mackenzie, Dorothy (1997) *Green Design*, London: Laurence King Publishing.

Pilatowicz, Grazyna (1995) *Eco-Interiors, a Guide to Environmentally Conscious Interior Design*. New York: John Wiley & Sons.

US Green Building Council (1996) *Sustainable Building Technical Manual*, Denver: Public Technology, Inc.

Zeiher, Laura (1996) *The Ecology of Architecture*, New York: Whitney Library of Design, BPI Communications.

Response

Re-manufactured sites?

Through several case studies, Dr Rogers offers a relatively simple but important observation. With the focus during the last several decades on the remediation of past contamination caused by industrial operations, Dr Rogers redirects our attention toward the future. She suggests that by considering the design of industrial and manufacturing facilities to be self-contained systems we may be able to prevent a repetition of this industrial legacy of environmental degradation. Dr Rogers suggests that this redesign includes the elimination of toxic materials during construction and on-site mechanical or biological treatment of contaminated by-products to eliminate their migration off-site.

This focus toward the design of future facilities is important, and Dr Rogers' rationale that environmentally sound practices are a "boon for business, design and the environment alike" has strong merits. Dr Rogers also demonstrates through the case studies that an evolving environmental ethic is transforming the design of many workplace environments. To offer a fair overview when discussing the operational and philosophical shift brought about during the 1980s and 1990s, one must also consider that many industries have chosen to relocate their operations to countries offering low wages and little in the way of environmental restrictions. Examples abound in Mexico, India and the Far East of industries dumping lethal toxins into the surrounding soil and waterways. From a global perspective, a discussion of industrial reform cannot just focus on the innovative, environmentally conscious European and American industries. They offer a valid representation of an economically viable restructuring from dirty to clean operations. But if the growth of industrial pollution is not to be spread along with production to other parts of the world, this model of clean design will need wider acceptance to compete. If proved economically viable, such a model would negate the perceived tension between economic growth and environmental protection in Third World countries. Critical to the validation of this model is economic analysis and further documentation. Until that time firms that implement environmentally sound design and operations need acknowledgement and support, and strict regulations will continue to play a powerful role in maintaining environmental performance levels for designers and operators. If other countries had equally strong standards the option of relocating and continuing damaging practices would be eliminated as an option, and we might well see more innovative designs to meet uniform standards for industrial development

In her paper, Dr Rogers makes a compelling argument that the price of remediation represents a direct increase to the cost of doing business, reducing the competitive edge for many industries. This argument would be strengthened if several examples with substantiated data were offered. Historically, wastes have been dumped with little regard for the long-term environmental impact, usually because it was the least expensive option. The area of cost comparisons is critical since proven economic analysis would offer strong incentives for a transformation in the design of industrial facilities and their operations.

There are several design implications to this integrative approach as suggested by Dr Rogers, including the planning of integrated cleansing and responsible material handling systems early in the programming stages of site development. What may be most provocative for designers, and not dissimilar to the design philosophy offered by Peter Latz in Chapter 11, is Dr Rogers' view that "technology is omnipresent on-site, an integral part of the infrastructure . . ." While different in some respects from the approach of Peter Latz, who preserves and readapts the industrial structures of the past, she does offer a similar vision through the integration of technology and natural systems. The resulting design implication will be the integration of artificial technologies with natural systems to treat the contaminated by-products. This dialectic offers provocative relationships and design challenges in which the esthetics of nature and industry may coexist.

Dr Rogers offers several case studies as examples of "green" buildings, worker-friendly environments, and a reduced use of toxic materials in the building design. However, in most industrial production, such as chemical manufacturing facilities, energy conversion plants or materials fabrication shops, the processes will still produce toxins as by-products. How in the future will designers and planners respond to these new site issues? One answer is that they already are responding. As environmental standards have become stricter, several changes such as on-site detention and retention of stormwater and liquid waste products must now be incorporated into the site planning. Subsurface collection systems, containment barriers and gas release systems are now found in both new sites and in older industrial sites being redeveloped. Maybe more interesting and certainly more speculative is how some of the technologies discussed in several of the papers offer new systems that will have design implications when planning for potentially hazardous conditions. For example, there are situations where designers might integrate phyto-mediation systems into the facility design for the treatment of spoils, see Chapter 4 for Eric Carman's "From Laboratory to Landscape." Will there be opportunities to create successionary forests, forests that are implemented sequentially as the spoils are amassed on-site? (See Chapter 13) Can bio-filtration be used to cleanse the wastewater on-site to reclaim the water for reuse in the operation? (See Chapter 12) It becomes clear, as it has in many of the papers, that an integrated planning team must be initiated early in the design process so that treatment options are considered to address on-site problems before remediation is required.

As was pointed out by McNeil and Lange in the essay in Chapter 6, "Engineering Urban Brownfield Development," historically industry has been tied to the urban sectors of this country. Today, with increasing "clean industries," including computer chip processing plants and high tech manufacturing, and the switch from railroad to air and trucking for product distribution, this historic tie to the city may be broken. Today many industries are relocating to rural and suburban locations. With this relocation, industries have greater quantities of surrounding land, land that can be designed for and used as on-site cleaning systems to detoxify waste products. The increased space offers the possibility to create large-scale wetlands and containment systems for storage of contaminated materials. The counterpoint is that so many of our lands have already been degraded, and the move to rural areas where land is relatively pristine to create new industries is a net loss of environmental quality. Would it not make better sense to clean up the old sites and recycle them for "clean industry"? Are there ways that industry can be rebuilt within our urban centers, utilizing whole systems design, or are the land parcels too small to integrate biological cleansing systems? As Dr Rogers points out, building recycling for readaptive uses, accountability of waste generation and recycling during building construction, plus sustainable landscape measures, are all important considerations when developing a new facility. The real legacy with which we are struggling today is the result of the inadequate disposal of toxic waste products, water and soil contamination from discharges and leakages, and airborne contamination from plant emissions. The greater question may be how, as we anticipate these occurrences, we can mediate them with systems that treat in place or contain toxins on-site for long-term treatment in a safe and sustainable manner. In addition to on-site measures Dr Rogers offers examples of both low-tech and high-tech solutions for responsible production. In the example of the Benetton corporation, using computer monitoring to link product need to availability reduced production, and the long-term storage on-site of potentially toxic materials became unnecessary. This relatively simple example indicates how efficiency of production design may reduce over-produced product disposal, with the energy consumption used for unnecessary production thus being eliminated. As one surveys the diverse range of technical and designed solutions for remediation presented in this book, it seems logical that through a forward thinking model these systems could be integrated during the development of industrial facilities and designed into the master planning, creating closed regenerative operations in which "enterprise [is an] ecological system."[1]

Note

1 Brozen, Kenneth, "Wilkhahn's Direction as a Design Leader," *Interiors* 148 (6) (New York: Billboard Publications, 1989, pp. 120–3).

Part III
Reclaimed landscapes

Part III comprises seven essays, six of which have supporting responses, that focus on design approaches to the reclamation of manufactured sites, including the research and application of advanced landscape technologies such as bioengineering, waste encapsulation methods and innovative ground-surface finishes.

The transformation of these sites involves decisions of what is to remain and what is to be removed – the importation, recycling or removal of site material, structures and utilities. Projects of national and international significance demonstrate the design influences on reclaimed landscapes, alongside descriptions of the conceptual design processes and their eventual development through to built work.

A broad introductory essay by landscape architect Rebecca Krinke places this work in both a historical and contemporary framework. The essay features case study projects of a number of contributors to this book presented in a recent exhibition at the Harvard Graduate School of Design. Krinke describes and analyzes these projects as celebrating and revealing past land uses through the clean-up strategies that were undertaken in their regeneration. Differing scales of manufactured sites and time periods over which work was carried out, and stormwater management and clean-up, are demonstrated in the large-scale reclamation of the 230-hectare Tyssen Steel works in Duisburg, Germany by landscape architect Peter Latz (Latz + Partner). By contrast, the Federal Medical Center project and the Robert Harbor Tidal Marsh, at 0.25 of an acre in size, presented by bio-engineer Wendi Goldsmith (The Bioengineering Group, Inc.), demonstrates an innovative new approach to the management of stormwater and site design proposals that integrate resource restoration with site planning.

William Young, a landscape architect formerly employed by the Department of Sanitation in New York, explains research, field testing and site design strategies developed on the Fresh Kills landfill restoration in Staten Island. This involves the integration of capping technologies, revegetation strategies, and habitat creation allied to landfill reclamation. Questions relating to what constitutes the correct restoration strategies for the landfill are raised. Kirt Rieder, a landscape architect, continues on this topic for the landscape reclamation project at Crissy Field, San Francisco, by raising and questioning "the issue of balancing current use with a proposed reuse of the site driven by historic precedent."

Finally, two essays explore differing aspects of the "Green Games" – the Olympic 2000 infrastructure and open space structure for Sydney, Australia. Michael Horne discusses examples of the integration of innovative construction technologies and design strategies and the initial research and development that was undertaken. Finally a detailed discussion on the open space structure focuses on the Northern Water Feature. Kevin Conger's essay explores the use of new technologies to innovate and express natural processes in conjunction with creative engineering and infrastructure. Together they demonstrate the viability of large-scale civic design projects to integrate and demonstrate science, engineering and design innovation from regional strategies down to the detailed site scale.

Chapter 10

Overview:

design practice and manufactured sites

Rebecca Krinke

Introduction

The projects in the "Manufactured Sites" exhibition highlight the role that technologies, employed to address a legacy of environmental degradation, have contributed to the built form of a newly designed landscape. Designers are increasingly asked to work on land with decades of environmental damage, oftentimes with multiple sources of contamination. Many potential development sites are within cities, where even though the land may have environmental problems, its location may make it too valuable to ignore. Environmental policies may dictate that remediation take place, and funding may be available to address the site's reuse. New paradigms of landscape types with their own environmental issues, such as landfills or military bases, may suggest new design solutions.

The case studies that are discussed in this chapter are located in the United States, Germany or Australia. The West is on the leading edge of remediating landscapes, although it is becoming a worldwide phenomenon. Western nations industrialized earlier and more relentlessly than other parts of the world, and differing forms of legislation have recently been enacted to address this environmental degradation.

The environmental legacy of the Industrial Revolution

The development of a more efficient steam engine in 1769, and the need to mine coal to provide fuel for it, were pivotal in launching the Industrial Revolution. Extracting and combusting coal poses severe environmental problems. Deep mines may acidify streams, and surface mines are difficult to be reclaimed for anything other than pastures (Hayes 2000: 29). Burning coal releases toxins into the atmosphere and produces acid rains that destroy immense areas of forests worldwide. The destruction of forests adversely affects air quality; plants cleanse the air of carbon dioxide and produce the earth's oxygen.

Oil and natural gas are the other carbon-based fuels that power industry. Oil spillage is a huge problem – not just the dramatic spills from ocean-going oil tankers, but the day-to-day spills and leaks from service stations and individual cars that account for millions of additional tons of oil polluting the water and soil. Natural gas burns cleaner than both coal and oil, but does release carbon dioxide into the air (Hayes 2000: 31). The nuclear power industry produces radioactive wastes that will not degrade to harmless levels for thousands of years.

Mining degrades the landscape in multiple ways. Extracting a desired metal from rock leaves behind vast piles of tailings (the excess material that did not contain the minerals), and the extraction process itself has its own environmental issues. For example, the gold mining industry has developed a technique that uses cyanide to extract gold from the ore.

Industrialization, including that of agricultural practices, has led to water and soil contaminated with chemicals. Contaminated soil and water introduce toxins into the web of plant and animal life. Pesticides in the groundwater means that sources of human drinking water are affected. Agriculture and other human development of the land have drained millions of acres of American wetlands. Wetlands are key to flood storage, erosion control, and the natural filtration of pollutants.

The earth's human inhabitants generate vast amounts of garbage, and disposing of this waste has become increasingly problematic. Western nations produce most of the world's garbage, and have most of the environmental restrictions on its disposal. In the recent past, garbage was often shipped to developing nations, promising environmental problems for the future of these countries. In the United States, most garbage ends up in landfills, although some of the nation's waste is recycled. If improperly constructed or damaged, contaminants can leach from landfills into adjacent soil and water.

The Industrial Revolution changed the human relationship with the earth. As the world moved from agriculture to industry, a mechanistic view of the universe began to supplant the idea of an organic nature. A desire for "progress" and faith in technology implied that the earth was a place to extract resources and its "complementary" idea: that the earth could absorb anything humankind asked of it. Land was typically viewed as a commodity, and in general this idea has prevailed until quite recently. It was in the 1960s when the environmental movement questioned the idea that the industrialized nations could continue with their same habits of production and consumption. The earth, water and skies were becoming visibly polluted. The first American environmental laws, such as the Clean Air Act of 1963, were passed. In 1970, the first Earth Day was held, and millions of citizens demonstrated their concern for the health of the planet.

The legacy of the Industrial Revolution – its disruption of the natural systems of air, water, soil, and habitat – is being addressed in the twenty-first century. Mandated by regulations, but also reflecting a change in society's values, despoiled sites are being reclaimed, and the "Manufactured Sites" case studies reveal this paradigm shift.

Design goals, theory and practice

Recovering a despoiled site involves the variables of the particular site (its problems and the program for its reuse), legal requirements that may have to be met, as well as the vision of the designer. The final design and form of the recovered landscape may be the visual and ecological integration of the project site into the landscape that surrounds it. This occurs primarily when the site has been damaged and the recovery restores the healthy functioning of the ecosystem.

Another approach is to recover the landscape with the intention of clearly revealing the hand of the designer, underscoring human intervention on the site. The site's history may be disclosed clearly or subtly alluded to, but restoring the landscape to a (perhaps unknown) "natural" appearance is not a goal of many of today's designers. Artist Robert Smithson's "earthworks" and their engagement of natural (including human) processes, and architect Peter Eisenman's idea of the site as a "palimpsest," have informed many of the "Manufactured Sites" projects.

It should be noted that the selection of Smithson and Eisenman does not mean to imply that these two individuals are the only operative influences on the case studies. Both of these theorists/artists were influenced by the written and built work of others. Smithson was part of the New York art world in the 1960s and 1970s and interacted with many groundbreaking artists and collectors. He was a voracious reader, with interests ranging from literary theory to geology to science fiction. Smithson's built work continues to inspire, but he may be even better known for his influential theoretical writings. He died in an airplane crash in 1973 while surveying a potential project site. Eisenman founded the Institute for Architecture and Urban Studies (IAUS) in New York (1967–85). He and his colleagues disseminated their ideas and the work of others, especially European architects, in print, exhibitions, and conferences. The IAUS publications were important venues for architectural theory. Eisenman continues to write and teach, and has his own architectural practice based in New York.

Robert Smithson and earthworks

Artist Robert Smithson's earthworks, sculpture, and writing in the 1960s and early 1970s prefigure many of the projects in the "Manufactured Sites" exhibition. His work offered a dramatically different attitude to art and the landscape than that which prevailed at the time. When the "earthwork" artists (Smithson, Michael Heizer, Walter De Maria, and others) began to create their art by directly engaging the earth as both the medium and setting of their work, they reawakened a dialogue between art and the landscape. These artists left their studios for the outdoors, primarily the deserts of the American West, to work with the particulars of a specific site. Their projects provided a counterpoint and critique of the art world's system of galleries and museums that showcased art as commodities

Smithson's work challenged the notion that culture and nature were opposites. He argued that human intervention on the earth is as much a part of natural processes as an earthquake. Smithson felt that it was inappropriate to disguise the post-industrial nature of a site and advocated that technology and human use is acknowledged. His work engaged time, change, and ambiguity. He issued a challenge to "explore the pre- and post-historic mind" (Holt 1979: 91). The "postmodern" sensibility that has emerged in the wake of Smithson has embraced ambiguity and multiple points of view.

Smithson's projects that engaged natural processes were influential in the art and design worlds. He was interested in a deliberate engagement of nature's processes of growth and decay. This was radically different from the landscape architecture of the time, which generally focused on exploring geometric form and solving functional problems. He wrote, "I'm interested in collaborating with entropy" (Holt 1979: 181), and this idea was exemplified in his project, Asphalt Rundown, where a truckload of asphalt was dropped down an already eroded hillside. "The 'first' erosion is underlined and made more permanent by the solidified asphalt, whose 'rundown' suggests an irreversible process of its own" (Shapiro [1995] 1997: 41).

The Spiral Jetty, Smithson's best-known project, was built in Utah's Great Salt Lake in 1970 near the site of an abandoned oilfield. The site's overlap of industry and beauty attracted Smithson. He engaged technology to build his project, while the elemental quality of the simple spiral recalls ancient monuments. Natural processes occurring on the site were intensified by the addition of the spiral of rocks. Salt crystals multiplied on the larger surface area, and, within the spiral, the indigenous pink algae flourished, intensifying the reddish color of the lake. Engaging entropy on this site had a dramatic effect: the earthwork has been underwater for much of its life due to the fluctuating water level of the lake.

Smithson's interest in the landscape extended to reading key pieces of landscape architectural theory. He commented on the writings of the most influential theorists of the eighteenth century: Uvedale Price, William Gilpin, and Edmund Burke. Price and Gilpin wrote about the "picturesque," which Smithson described as a synthesis of the "beautiful" (characterized as symmetrical, self-contained) and the "sublime" (characterized by vastness, awe inspiring, verging on terrifying). The "picturesque" was described by Price and discussed by Smithson as the introduction of time into one's experience of the landscape. Price used the example of a gash on the earth, which as it is softened over time by vegetation becomes "picturesque."

Smithson admired landscape architect Frederick Law Olmsted as "America's first 'earthwork' artist." In his essay "Frederick Law Olmsted and the Dialectical Landscape," Smithson praised Olmsted for understanding and utilizing the theories of Price and Gilpin. Smithson discusses Olmsted's design for New York's Central Park (1858) and lauds the fact that the park is completely public, unlike most models of the time.

When Olmsted began work on Central Park, the site was virtually devoid of vegetation and strewn with rubbish. Smithson admired Olmsted's vision to both work with the site itself (for example, the glacial outcrops that were incorporated into the design), and to see the park as a "dialectic between the sylvan and the industrial" (Holt 1979: 121). The park was a human-made construction, not a remnant of untouched nature as many people still believe. It is a powerful work of art, and, as Smithson underscored, it is a work of art that is ongoing.

Smithson wrote, "The authentic artist cannot turn his back on the contradictions that inhabit our landscapes" (Holt 1979: 123). He believed that art could be a mediating force between ecology and industry. He wrote to mining companies recommending that artists be included in the process to reclaim mines. Before his untimely death, Smithson was involved in negotiations with the Minerals Engineering Company of Denver. He proposed reshaping their mining tailings into earthworks, not disguising the nature of the site but adding an esthetic dimension to it. His plans included strategies for locating millions of tons of tailings that would be deposited over time (Beardsley 1989: 23).

Smithson's interest in despoiled landscapes has expanded notions of what is viewed as beautiful. To many contemporary designers, mines, urban wastelands and other damaged sites have a unique beauty – one that is filled with potential. His work has inspired countless artists and designers to engage nature as a process, rather than as scenery.

Peter Eisenman and "palimpsests"

Thinking of the site as a "palimpsest" is a design idea that emerged in the 1980s in the writings of Peter Eisenman. A palimpsest is a manuscript or tablet that has been written upon several times, often with earlier, imperfectly erased writing still visible. Thinking of the site as a palimpsest allows designers to utilize the site's layers of history to reveal aspects of the site, or even to add a new layer of self-conscious fiction. Using principles of collage and juxtaposition, history is seen not as linear phenomena, but as layers or discrepancies between a past event (history) and present recall (memory) (Nesbitt 1996: 175).

The theories of the French philosopher Jacques Derrida have informed Eisenman's work. Derrida invented the term "deconstruction" to describe his method of asking philosophical questions of texts. This reading of the text is designed to "reveal logical or rhetorical incompatibilities between the explicit and implicit planes of discourse" (Groden and Kreiswirth 1994: 185). "Much is made in Derridean deconstruction and in Eisenman's version of it, of uncovering the hidden, of revealing the repressed" (Ghirardo 1990: 84).

In his design for the Wexner Center for the Visual Arts at Ohio State University, Eisenman makes visible his notion of the site as palimpsest. The team of Eisenman/Trott in joint venture won an invited design competition for the Center; the project was completed in 1989. Laurie Olin, principal of Hanna/Olin Ltd, was the landscape architect. The siting of the building was to a large extent up to the competitors; the teams were shown an area for the Center to be located rather than a prescribed site. The Eisenman scheme was the most radical in its siting and the most experimental in its form.

The Wexner Center occupies an interstitial space between two existing buildings, and is more like a passageway or a landscape than a building. Much of the building is below grade. The project is not readily distinguishable as a mass; it is more like a set of relation-ships with the site (on several scales), with the ground plane, with interiors and exteriors, and with time.

The Wexner Center is located at the city/campus interface. The project subsumes the path from the main entrance to the campus to the primary green space, The Oval. As Eisenman investigated the site he discovered that two different grids were active on the site. The grid of the nearby city streets and the grid of the campus were different by 11 degrees. The city grid enters the campus at 15th Avenue, grazes one edge of The Oval, until it is picked up again at the far edge of the campus where it aligns with the football stadium. Extending the city grid into the campus grid through the Wexner Center makes these two different grids visible, and is a primary way that the site as a palimpsest is expressed.

The grids operate in two and three dimensions, blurring distinctions between floor, wall, and ceiling as well as notions of inside and outside. The traditional idea of the grid as a system of order and structure is overthrown in the Wexner Center by grids that are juxta-posed and overlapped at different scales. An immense gridded white lattice, called the "scaffolding," is the device that denotes entry and circulation within the project. The scaf-folding interacts with the different types of glass that make up the building walls, multi-plying and changing the effects of the grid with differing effects of light and shadow. Grids are also expressed within the building; for example, the ceiling and lighting of the below-grade theater utilizes both grids that are operative above grade. Outdoors, a grid of linden trees overlaid with a different grid of ginkgo trees seeks another relationship between the city, the Center and The Oval.

The ground plane is also filled with ambiguities between systems of measure and order. The terrain tilts and gives clues to spaces in the building below. Raised planting "plinths" filled with indigenous plantings suggest the native prairie, Indian burial mounds, and deeper geological substrates. Walking between the plinths may transport one to day-dreams of being below ground. At one point, the plinths are interrupted to make visible the "Greenville Trace": the line across Ohio where two different surveys failed to meet. This connects the project to larger systems of survey grids that overlay the country.

A second way that the palimpsest is expressed in the Wexner Center is via the Armory towers. Ohio State's Armory was previously located on the site of the Center, but was demolished in 1958 after a fire. Eisenman's scheme resurrects the Armory as sculpture. His work with the Armory challenges the idea of a site as place-, time-, and scale-specific. The Armory is recalled, not rebuilt, on the site in various ways. The towers are in different posi-tions and built to a much smaller scale than the original Armory. Depending on where one is looking from, the towers may appear as solid forms or as fragments. The brick expres-sion within the towers varies: some portions have low relief grid patterns, which links them to other grids in the project. The technique employed to construct the towers of bricks make it appear that the towers are actually in the process of demolition.

The foundations of the Armory are expressed in their actual location on the site with lines of pavement and walls. Ironically, portions of the original foundation were present on-site when construction of the Center began, but they were inadvertently destroyed and were rebuilt. Resurrecting the Armory may be a way of revealing the "repressed text" of the site. Armories are military installations on college campuses and there has been a recent history of uneasy relationships between the two institutions (Ghirardo 1990: 84).

Eisenman's Wexner Center for the Arts challenges architectural convention in several ways. The project did not employ a holistic design strategy or provide one strong memo-rable image. The idea of the palimpsest provides a way to analyze a site and create a project from multiple, shifting vantage points. His work has been important to many designers, as will be seen in the case study portion of this chapter.

Historical precedents

Reclaiming a site that has been damaged or has outlived its original use is not a new concept. The Back Bay Fens of Boston remade the landscape for flood control, and the Parc des Buttes-Chaumont in Paris recovered an abandoned quarry as a public park. Both of these projects were built in the late 1800s and are important precedents for the "Manufactured Sites" exhibition projects.

The Back Bay Fens

Frederick Law Olmsted, considered the founder of landscape architecture in America, began designing the Boston Parks system in 1878. Boston at this time was already a city radically reshaped by human intervention. By as early as 1645, every marsh within the city boundaries had been altered in some way (Zaitzevsky [1982] 1992: 9). The city's hills had been cut down, and this soil was used to extend the city's landmass into the ocean.

The Boston parks system, built over twenty years from 1878 to 1898, was the largest construction project in the history of the city (Zaitzevsky [1982] 1992: 151). A key piece of Olmsted's strategy for Boston was the Back Bay Fens. Back Bay is a fashionable residential district that was created by filling in a portion of the Charles River estuary and adjacent marshes. A park was desirable in this area for two reasons: there were drainage problems that had to be addressed and property owners and developers knew that a park would increase the market value of their land. Olmsted saw the Fens primarily as a sanitary improvement, and not a conventional park, as the project's main objectives were to provide flood control and sewerage. The Charles River was connected to the ocean and therefore tidal. Stony Brook and the Muddy River both drained into the Charles. When there were extremely high tides, these two streams flooded. Compounding this problem, raw sewage was dumped into both streams, and eventually reached the Charles. This pollution was not only unsightly and foul smelling, it proved deadly to the last vestiges of remaining salt marsh.

Before Olmsted was hired, the city engineers had made a preliminary proposal to address the flood control problem. They proposed a large rectangular masonry storage basin with a tidal gate to control the water level. Olmsted felt this solution would be both unattractive and expensive. He was interested in solving the flood control problem through regrading the site to contain the flooding waters and restoring the salt marsh, an ecological community that is adapted to daily tidal inundation.

Olmsted's plans for the Fens called for intercepting sewers to be constructed to solve the sewage problem. A covered conduit was employed to divert the Muddy River into the Charles. The regular flow of Stony Brook was also diverted this way. It was the floodwaters of Stony Brook that specifically had to be contained. The two streams were diverted to minimize the freshwater impact on the salt marshes that he planned to restore on the site.

Olmsted worked closely with City engineers to develop his project. He employed their ideas of a tidal gate: when the water level in Stony Brook rose above the level of water in the Fens, a pair of gates would open into the Fens.

> By flooding the salt marsh, the 30 acres normally covered by water could be increased to 50 acres without raising the water level more than a few feet. As the tide dropped, the extra water would empty out into the Charles over a dam under Beacon Street. Gates in the dam regulated the tide in the Fens, which normally rose only a foot between low and high water. The gate chamber at Stony Brook also included a sluiceway through which larger amounts of salt water could be admitted periodically at high tide through the Stony Brook conduit to prevent stagnation   (Zaitzevsky [1982] 1992: 154).

The Fens project was an experiment in planting design. Olmsted scholar Cynthia

Zaitzevsky notes, "This choice [restoring salt marshes on the Fens site] committed him to a planting program so innovative that it literally had no precedent" (Zaitzevsky [1982] 1992: 186). Olmsted's restoration differed from a naturally occurring salt marsh in several ways. A salt marsh found in nature would have mud banks exposed during periods of low tide. The water level in the marsh that Olmsted designed would fluctuate by only one foot, thereby decreasing the appearance of bare slopes. Naturally occurring salt marshes contain plant communities that are associated with specific water levels and their associated salt content. In Olmsted's salt marsh design, with its smaller vertical change in tidal fluctuation, his planting palette could be both more condensed and more extensive than would occur in nature (Zaitzevsky [1982] 1992: 186).

As Olmsted wrote:

> A mere imitation of nature, however successful, is not art, and the purpose to imitate nature, or to produce an effect which shall seem to be natural and interesting, is not sufficient for the duty before us. A scene in nature is made up of various parts; each part has its individual character and its possible ideal. It is unlikely that accident should bring together the best possible ideals of each separate part . . . and it is still more unlikely that accident should group a number of these possible ideals in such a way that not only one or two but that all should be harmoniously related one to the other. It is evident, however, that an attempt to accomplish this artificially is not impossible . . . The result would be a work of art . . .
>
> (Sutton [1971] 1997: 15)

The transitional areas of groundcovers and shrubs were selected for their salt tolerance. Only where the Fens met the city edge was there an edge of turf with street trees. Olmsted wanted visitors to the Fens to think that the city had simply grown up around the marsh. He wrote that the effect of this would be "novel," but also that the salt marsh would possibly be found more poetic for "town-weary minds than an elaborate and elegant garden-like work" (Zaitzevsky [1982] 1992: 57).

Today, the salt marshes are gone from the Fens. With the Charles River Dam's construction in 1910, the Charles is no longer tidal. Olmsted's office made a proposal to redesign the Fens to function without salt water, but nothing was done, and in fact, as a freshwater corridor, the Fens became heavily polluted with trash.

Olmsted titled his design for the Fens a "Sanitary Improvement or the Back Bay Drainage Basins," rather than a park. He emphasized that the design solution was to provide flood control and sewerage, which in turn increased public health benefits. This concept links his work to one of the reasons that "manufactured sites" need to be reclaimed today: they are places of public health concern. As in many contemporary projects, Olmsted completely remade the site: the Back Bay Fens was created by extensive filling and dredging. The earthwork that he designed to contain floodwaters was a marriage of esthetics and function that prefigures much of the design work discussed in this chapter.

Parc des Buttes-Chaumont

In Parc des Buttes-Chaumont, completed in Paris in 1867, quarries that became dumps were recovered as a highly artificial public park. As Paris grew, this industrial waste site originally on the outskirts of the city was too important to be neglected. In the nineteenth century, parks had become public. They were seen as important contributors to the physical and social development of citizens, and as a way to showcase cultural achievements. The opening of Parc des Buttes-Chaumont was planned to coincide with the 1867 World Exposition, Napoleon III's celebration of French innovation in industry and the arts (Meyer 1991: 19).

Parc des Buttes-Chaumont was designed by J.C. Adolphe Alphand; trained as a civil engineer he designed many parks, squares and boulevards in Paris in the late 1800s. Echoing the philosophy of Olmsted, Alphand believed that the garden was a work of art and should not attempt to copy nature. He too was interested in a conversation between the human-made and the organic. In Parc des Buttes-Chaumont, Alphand embraced the challenges of an oddly shaped parcel of degraded land crossed by a railway line. His park is a masterful work that unites art and engineering.

The park's infrastructure and other functional requirements were viewed by Alphand as opportunities. He wrote:

> Through the artifice of embankments and landscaping, one may even gain advantage from a big road or railway running alongside or through a park. If the work is well done, the trains, carriages and pedestrians which seem to circulate within it, give a degree of animation, and the talent of the designer transforms what seems to be an irredeemable fault into positive ornament . . . One can also turn the view of a station, and industrial building or any construction to advantage, providing that it is fittingly framed in the landscape.
>
> (Vernes 1984: 60, fn.)

The site's unusual, almost violent topography was exploited in Alphand's design. He made the central spire of rock larger and capped it with a temple, where a stairway descends to the water or to a dizzying suspension bridge. Quarry walls were reshaped, and the remaining topography was regraded for a smooth and grassy contrast to the rocky focal points. The site's topography and the park's paths were united by a synthesis of esthetics and function. Alphand's objective was to link focal points, such as park entrances and buildings, via a consistent curve (never with straight or serpentine lines). This strategy provides a fluid, changing view for the pedestrian and the best fit to the evenly shaped contours.

The park provides a fascinating contrast between highlighting actual geology and the manufacture of an artificial geology. Sophisticated engineering technologies were employed to create an artificial lake and waterfalls. An innovative application of concrete, was employed to produce artificial rocks. The fabricated geology extended to creating concrete stalactites and stalagmites in the underground cavern. Throughout the park, the fantasy continues with concrete steps and rails made in the form of logs.

The artificiality of the park makes it a theatrical space of exploration and promenade. Louis Aragon, a key member of the Surrealist movement in art, wrote about a nighttime visit to Parc des Buttes-Chaumont in his 1926 book *Paris Peasant*:

> The Buttes-Chaumont stirred a mirage in us, one with all the tangibility of these phenomena, a shared mirage over which we all felt we had the same hold . . . At last we were going to destroy boredom, a miraculous hunt opened up before us, a field of experiment where it was unthinkable that we should not receive countless surprises and who knows? a great revelation that might transform life and destiny.
>
> (Aragon [1926] 1994: 133)

The Parc des Buttes-Chaumont was created from a site scarred by industrial processes. Alphand exploited the quarrying of the site in his design; the history of the site was not hidden. Geology both real and artificial was celebrated. Employing the technological processes and materials of the time, he created a park with layers of real (and invented) natural and cultural history. Alphand's Parc des Buttes-Chaumont provides an example from the nineteenth century of the site as a palimpsest.

Introduction to the case studies

The projects that are described below were all featured in the "Manufactured Sites" exhibition. When the projects have addressed similar issues, such as restoring a marsh, they have been grouped for comparative purposes.

Restoration

Saw Mill Creek Salt Marsh Restoration, New York
Crissy Field, San Francisco, California

The Saw Mill Creek Salt Marsh Restoration and the Crissy Field project both involved the creation of salt (tidal) marshes, although for different reasons and to different effects. The Saw Mill Creek Salt Marsh project utilized restoration and bioengineering to repair the effects of a toxic oil spill. At Crissy Field (which is under construction at the time of writing) a salt marsh is being created where one once was. The designers' goal at Crissy Field is not to restore the marshes to their original configuration, but rather to create a marsh that makes it very apparent that these marshes are the work of human intervention.

In the Saw Mill Creek project a disastrous oil spill devastated the existing salt marshes. The priority in this project was restore the marsh, and another goal was to assess whether the reintroduction of *Spartina alterniflora* (salt marsh cord grass) could reduce total petroleum hydrocarbon (TPH) levels in the soil. Crissy Field, a National Park Service project, asked the designers, Hargreaves Associates, to recreate the salt marshes that once existed on the site as part of their design. A re-creation of a historic airfield was also mandated. These two requirements for the park showcase compelling issues of what historical period a site should be restored to and how restoration goals that may compete historically can be resolved – i.e., how would two things that existed in the same location, but at different times, both be restored?

The **Saw Mill Creek** project was necessitated by a 1990 oil spill of over half a million barrels of fuel oil into the Arthur Kill, a tidal strait separating the western shore of Staten Island from New Jersey. While tidal marshes were once viewed as swampy wastelands, they are now known to be critical for facilitating the ecological processes of habitat creation, water filtration, and erosion control. The oil spill destroyed birds, fish, and smaller animals, as well as hundreds of acres of low marsh vegetation consisting primarily of *Spartina alterniflora*. Exxon was found legally responsible for the spill and, through the damages awarded, the City of New York created a Salt Marsh Restoration Team within their Natural Resources Group (Marton 1998: 84).

The effects of fuel oil on salt marshes have been documented since the 1970s, but no attempt had been made to restore vegetation to areas impacted by oil. The efforts along the Arthur Kill therefore represent an experiment that has implications beyond the scope of the January 1990 spill and subsequent restoration. The technology to restore the marsh was extremely simple, but laborious: hundreds of thousands of plugs of *Spartina alterniflora* have been planted by hand. Hundreds of volunteers collected more than six million seeds, replanted them in nurseries, then transplanted the shoots to the marsh. The addition of one ounce, time-release fertilizer to each plug proved indispensable in achieving the goals of oxygenating the cord grass root zone as well as the beneficial bacteria found in the root zone. This bacteria uses carbon in the fuel oil for its own growth, degrading the oil and bringing down the level of total petroleum hydrocarbons in the process. On average, fertilized plants were more than twice as likely to survive (Marton 1998: 87).

This project highlights the need for restoration (in cases like the Arthur Kill oil spill) in two key ways: areas that were not replanted have not been naturally recolonized by cord grass, and the replanted areas now have lower concentrations of oil than areas that

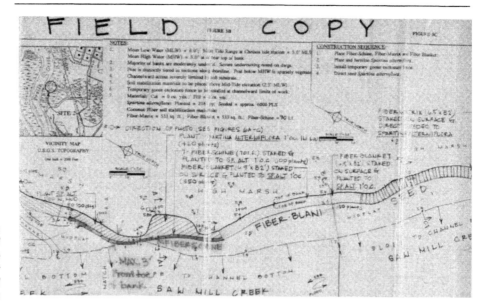

Figure 10.1 Salt Marsh
Restoration Team field
notes show areas to be
seeded, fenced, or
supplied with soil
stabilization materials

weren't planted. This process, whereby plants and bacteria assist in cleansing contaminated soil, is known as bioremediation. This project was crucial to proving that oil spills can be healed with biological methods.

Landscape architects worked with scientists to apply scientific research to an actual site, creating the field notes as shown in Figure 10.1 and specifications that translated research into a restored marsh. A key aspect of this restoration, and what sets it apart from most others, is its thorough monitoring program. As with the planting, volunteers assist with this monitoring. Results have been carefully documented, expanding on available scientific research. This salt marsh restoration by the City of New York has educated countless citizens on the importance of salt marshes to the ecology of the area. Now, ten years after the oil spill, the restoration is a spectacular success. Success in this project has been defined by the Restoration Team in two ways: (1) restoring the plant and animal life of the marshes, and (2) the restored marshes are visually indistinguishable from undamaged marshes.

It is interesting to note that the Arthur Kill is a heavily trafficked shipping lane. The wave action from boats was a concern in the establishment of *Spartina* seedlings. Some transplants were protected by wave dispersion devices, but seedlings that were not protected survived just as well. There are indications that areas of very high wave energy may require permanent wave barriers to persist over the long term (five years and beyond) (Marton 1998: 87). The design and placement of these barriers could add a fascinating, sculptural layer of physical form to the science of tidal marsh creation.

Crissy Field represents a new typology of sites that need recovery: the closure of military bases in the United States. The soil and groundwater at these sites is typically contaminated and needs to be remediated. The sites also need to be redesigned for their non-military life.

Crissy Field is a public/private partnership between the National Park Service and a non-profit, fund-raising organization, the Golden Gate National Parks Association, which retained Hargreaves Associates to design the new Crissy Field. The National Park Service has several sections, including a cultural resources, a natural resources and an interpretative division. The transformation of Crissy Field from military use to public open space involved discussion with all three of these groups. Clarifying and reconciling the goals of these groups was part of the work that Hargreaves Associates did on the Crissy Field project.

The work of Hargreaves Associates is one where the hand of the designer is very visible. Expressing the overlay between natural and cultural systems is something that the firm is known for. Crissy Field is in San Francisco, a dense urban area with a long history of settlement; the site has been reworked many times. In the redesign of Crissy Field, Hargreaves Associates effectively engaged the idea of the site as a palimpsest. The Crissy Field design engages different times in history to achieve both of the National Park Service goals: restoring the tidal marshes and the historically important airfield.

The original 180 acres of marsh and the airfield cannot both be accommodated on the site: it is simply not large enough and there are other goals for the site, including parking lots. Hargreaves Associates employed hydrologists, marine engineers, and planting consultants to assist them with the establishment of the salt marsh on the site. The hydrologists calculated that twenty acres was the minimum amount of area necessary for the marsh to be self-sustaining. The project restores this acreage in the center of where the marsh was historically. The technical constraints of constructing the marsh were fundamental to its final form, which in turn influenced the form of the entire park. The key determinant was the optimum location for the tidal channel, which will provide the critical exchange of water that must occur twice a day to ensure the health of the marsh. The new marsh is clearly constructed, characterized by sculptural landforms, as shown in Figure 10.2. The park's design restores the airfield in its original location, in its exact size and shape, although it too is configured as a landform, underscoring its remade nature.

The interpretive layer of the park functions "in the present tense" as the site's history will be discussed through exhibits and tours. Crissy Field also engages another layer of

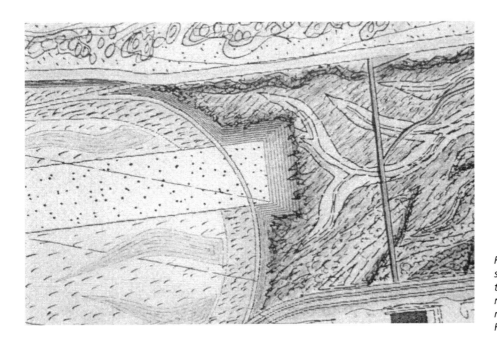

Figure 10.2 Detail of site plan highlighting the interface between restored airfield and restored marsh at Crissy Field

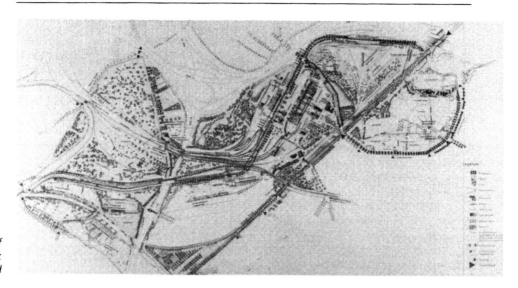

Figure 10.3 Site plan of
Landscape Park
Duisburg-Nord

time: the new tidal marsh is designed to transform incrementally from exposed mud flats to salt marsh cord grass (*Spartina alterniflora*) over ten years.

Reuse/landscape of memory

Landscape Park Duisburg-Nord, Duisburg, Germany
Landscape Park Duisburg-Nord is one of nearly a hundred projects in the International Building Exhibition Emscher Park (IBA) in the Ruhr District of Germany, a setting of 2.5 million inhabitants. A primary goal of the IBA is the ecological renewal of old industrial areas to promote new economic development. A key piece of the IBA is the Landscape Park Duisburg-Nord, as shown in Figure 10.3, where a blast furnace plant is undergoing a metamorphosis into a park. Peter Latz of Latz + Partner, the landscape architectural firm based in Kranzberg, Germany that designed the park, describes their intentions:

> The tasks of dealing with run down industrial areas and open cast mines require a new method that accepts their physical qualities, also their destroyed nature and topography. This new vision should not be one of "recultivation", for this approach negates the qualities that they currently possess and destroys them for a second time.
>
> (Latz 1998: 17)

The blast furnace, an artifact of twentieth-century industrial processes, provides a stunning sculptural presence in the new park. The site is a complex matrix of buildings and landscapes, and the designers' goal was to utilize the existing fragments of industry as layers that are recombined through the lens of park design. Latz explains:

> These layers connect only at certain points through specific visual, functional or merely imaginary linking elements. The uppermost layer is the railway park with high level promenades and the lowest layer is the deep-set water park. Other individual systems are the connecting promenades at street level, or the single fields and clumps of vegetation. All these different layers or single parks are connected by the system of linking elements – either symbolically by gardens or substantially by ramps, stairs, and terraces.
>
> (Latz 1998: 6)

The blast furnace with its haunting visual qualities is an ever-present reminder of the site's industrial past. The realities of this past include contaminated soil and water. Less contaminated areas, and other large parts of the 200-hectare park, are left to develop on their own without intensive treatment. For example, birches are colonizing the black waste material of the coal washing process. This coal mine spoil covers a former coking plant that is heavily contaminated with polyaromatic hydrocarbons (PAH). Two alternatives were evaluated to deal with this problem: capping the hazardous materials with layers of clay (with no planting on top of the cap), or in the alternative that was selected – accepting a slight gas diffusion over the course of several generations, slowly reducing contamination, while allowing for limited use such as walking or bicycling. There is a dialogue between the planting allowed to evolve according to natural processes and management interventions and maintenance. In the most intensively treated and used parts of the park there are hedges and grids of trees that are carefully maintained.

Heavily contaminated portions of the park are closed to visitors and unused. Toxic materials, such as some of the demolished buildings, were removed and deeply buried within the ore bunkers; they are sealed and covered with "roof gardens" as shown in Figure 10.4.

Figure 10.4 "Roof gardens" within Landscape Park Duisburg-Nord were made by sealing toxic materials below

Within the railway area of the park, high dams consisting of slags of a high pH value were created to immobilize heavy metals. These new landforms may be walked upon by visitors. The paths are covered with dolomitic limestone chips, a material that also has a high pH value. This new topography is higher than any surrounding vantage point and offers unique views of the park and its setting.

Within the park, the processes of decay, dismantling, repairing, and constructing are all active. The old industrial infrastructure: buildings, bridges, railroads, tunnels, ore deposits areas, and parking lots are reinterpreted as a park. The goal of the park design is not to rip out the old and create a new "blank slate" but rather to acknowledge and celebrate the site's history. This remembrance is not via preservation of the blast furnace as a museum, but, rather, the buildings have been given an active new life. Only the dangerous portions of the structures have been strategically removed. The new life of the buildings and landscapes takes place as an overlay to the existing structure of the site. For example, the central power station, which has space for 8,000 people, is used for performances and exhibitions. Smaller engine rooms hold concerts and receptions. The structures are not remodeled; the new uses take place within the old settings.

Recycling is a major task on the site: the remnants of demolished buildings are ground and transformed into new soils for planting, into roads and walks, and into concrete used in the park. New catwalks were added made from recycled materials found on the site. The catwalks have been painted in bright colors to invite exploration by visitors, in contrast to the untouched gray and rusty portions of infrastructure. Staircases were developed from dismantled maintenance walkways and from old crane bridges.

The water park layer of the design remodels the old Emscher, an open wastewater ditch which carried non-clarified sewage to the Rhine River. Latz + Partner used the profile of this old construction to avoid contact with the contaminated soil that surrounded it. Within this profile, a new clear water system has been installed, and the wastewater is carried out in a 3.5 meter diameter underground main. The open canal now runs clean with water collected and purified from buildings and pavements of the site. Rainwater flows in open rivulets and through the existing overhead pipe systems, where it falls into the former cooling basins, becoming enriched and purified by oxygen. These open and visible water systems are planned to encompass much of the whole blast furnace site. The natural processes of water, united with technology, invite the visitor to see both with new eyes.

Different layers of time are simultaneously coexisting in the Landscape Park Duisburg-Nord. The park is conceived of as an ongoing process; it is alive and complex. The old industry provides an armature for new experiences of scenery and activity. The design of the park acknowledges that many people are attracted to unconventional settings. The park has served as the backdrop for films and concerts, creating a new layer of memories on the site. The huge ore bunker walls have been claimed as a "climbing garden" by the German Alps Club. This is only one example of the many different clubs and individuals working in the park. There are guided tours in several languages, and self-guided tours through the park that reveal and interpret the site's industrial history.

Latz + Partner's park is a powerful vision of the reusing and remembering of a landscape. Conceived of in layers, both spatially and historically, Landscape Park Duisburg-Nord has a correspondence with the Parc des Buttes-Chaumont and the contemporary interest in exploring the site as a palimpsest. Landscape Park Duisburg-Nord combines human intervention and natural processes to create an environment that neither could have created alone. And the creation of this park is ongoing.

138

Wasteland/public park

Byxbee Park, Palo Alto, California
Spectacle Island, Boston Harbor, Massachusetts
Stockley Park, London, England

Three landfill projects were highlighted in the exhibition: Byxbee Park, Spectacle Island, and Stockley Park. Byxbee Park embodies qualities that are now associated with parks on landfills, such as the absence of vegetation. At Byxbee Park, highly sculpted landforms dramatically reveal that the site is human made. The capping of Spectacle Island is unusual in that it provides generous depths of loam for the planting of trees and shrubs. This topsoil was created through a new technology, the "manufacturing of soil." Stockley Park highlights an emerging area of site recovery: this project moved an existing landfill. Landfills are relatively new, and the fact that they are already being moved demonstrates that the environment will be unceasingly remade and that technologies will constantly emerge to facilitate this remaking.

Byxbee Park, by Hargreaves Associates, is a work of art on and about a landfill. It is a highly artificial, sculptural landscape. The form of the park is shaped by esthetic intent, and informed by the constraints of building atop a landfill. Native grasses are used; no irrigation is provided. The park is bare of all shrubs or trees due both to the shallow soil depth (12 inches) over the cap, and the fact that roots may break the cap containing the landfill. The paths of crushed oyster shells are permeable.

Composed of earth and grasses, and punctuated by objects and landforms that are inspired by the nature and culture of the site and its surroundings, the artificial form of the park explores the site as a palimpsest. Paths wind through the park providing views of a series of events/places/follies such as a field of poles as shown in Figure 10.5, a flare

Figure 10.5 A field of poles at Byxbee Park highlights the landfill's constructed nature

(shaped in plan like a keyhole) that burns off excess methane gas, and a chevron of jersey barriers, among others. The poles conjure up images of the large fences that typically surround active landfills, and link the park to the surrounding highways; they also speak to minimal art. The flare provides a clue (or a confirmation) that the site is a landfill. Connecting the site with the larger context is the chevron of jersey barriers which line up with the flight path to the local airport, arranged in such a way that signals (via aeronautical symbols) "don't land here." These views along a path link this park to nineteenth-century picturesque parks, while the highly sculpted landforms link it to the earthworks art of Robert Smithson, as well as to more ancient forms such as burial mounds.

The park is elevated and enlivened by mystery, and reminds one of the role of mystery in art. It has a primitive and elemental feeling. The sign at the entrance is a simple metal bas relief of the park's topographic features, almost as if the sign is to be read in to the distant past or future when written language itself may be different, or lost entirely. Byxbee Park doesn't tell the visitor what to do by providing familiar places such as tennis courts. As parks upon landfills become more common, visitors may understand that this park sits atop garbage. At the very least, the visitor may wonder why this park is not like other parks, and may begin to investigate this question by exploring the site on a deeper level. The park design is strategically engaged in both revealing and concealing.

Byxbee Park is physically one of clarity and simplicity, where earth and sky predominate and merge, and a human being is the mediating element between the two. The cleanness and emptiness of this sculpted park is an inspiring and elevating change from the strip mall and freeway banality that one must pass through to get to it. The clarity of seeing powerful forms without plants is unique and refreshing. Certainly the park feels somewhat raw. The native grasses have a texture and presence that is more aggressive than a mown lawn. The park has a quality of the primordial.

The idea of park merges with the reality of landfill in Byxbee Park. The landfill inspires the park and the technological constraints take the park to a new level of form. It does not try to blend into the surroundings or be a conventional park. (Most parks on landfills are designed for active recreation: soccer fields, ball fields, and golf courses.) Byxbee Park does more than acknowledge that it is a capped landfill: it celebrates this moment in time when parks are made atop landfills.

Spectacle Island, one of Boston's many Harbor Islands, has recently undergone a radical transformation: the island has been reconstructed as a landfill. While Byxbee Park was designed on top of a capped landfill, at Spectacle Island, the island's form was determined by the need to dispose of the maximum amount of fill possible and the stabilization of that fill. The depression of Boston's Central Artery (the new tunnels through downtown that are being dug to contain Interstate 93) is creating huge amounts of fill that have been barged to Spectacle Island.

Spectacle Island has historically been used as a site for dumping; capping the island has been an environmental priority for some time. When the Massachusetts Highway Department began work on Spectacle Island, methane gas produced by decomposing garbage was burning, cliffs of refuse were eroding and falling into the harbor, and the beaches were strewn with litter.

Spectacle Island has natural dramatic landforms (drumlins that were formed 10,000 years ago by the receding of the glaciers). This slow geological time and the resultant form may be contrasted with how Spectacle Island has been reconstructed. The huge amounts of fill have dramatically increased the scale of the drumlins – doubling their height in the space of just a few years. The island's new form, as shown in Figure 10.6, makes reference to drumlins, but it is recognizably human made.

The overall design intent of the portion of the Spectacle Island project displayed in the "Manufactured Sites" exhibition is the protection of the physical integrity of the island's

Figure 10.6 View of Spectacle Island shows the island being remade with fill barged from the Central Artery project in Boston

cap by providing new soil that will stay in place. The landscape architectural firm of Brown and Rowe designed this phase of the project. (The island will eventually become a park. This is a separate project to begin at a future date.) A topsoil base of 4–5 feet provides planting depth for thousands of trees and shrubs, preventing erosion and decay of the cap. In order to provide the enormous amount of topsoil required without stripping valuable local topsoil, a new technology – the "manufacture" of soil – was determined to be the best solution to the topsoil requirements. Brown and Rowe worked with soil scientist Dr Philip Craul to develop the optimum mix. Materials available right on the island (glacial till) and other materials (such as sand), some recycled (compost), barged over from the mainland are being combined to create topsoil.

Meeting specific grading criteria and selecting plantings for their erosion control abilities (native, fast-growing plants such as sumac, blackberries, blueberries, viburnums, and clematis) are key to maintaining the soil's stability. Spectacle Island's long-term success as a landfill and a park depends on this stabilization. The establishment of a sustainable successional landscape is necessary so that the soils can withstand extraordinarily severe environmental conditions, including strong waves and winds. There is no source of fresh water on the site other than rainfall. There will be little maintenance after the completion of construction.

The pattern of planting relates directly to topography. Multi-layered vegetation in protective linear bands parallel to topographic lines will intercept run-off and diffuse winds. The plant species are a mix of invasive pioneers and successional plants, which will progress to a mature coastal woodland community. Shrubs are underplanted with non-competitive grasses and legumes that will increase the water-holding capacity of soil for the first two years until the shrub roots knit together. Deciduous trees are underplanted with grasses and legumes to hold slopes until woodland canopies fill in.

The vegetation pattern will evolve differently from what is placed during construction. Meadow mixes on slopes will typically not be mown in order to encourage the growth of shrub and tree seedlings. Resilient trees such as black locusts, which can regenerate even after being blown down in a hurricane, are planted to increase the long-term success of the island's plantings. An important criteria for plant selection is the presence of fruits, nuts, and berries to attract wildlife. Both birds and mammals spread seeds, which assists in revegetating the island. A new ecosystem maintained by natural plant and wildlife processes will adapt to the site's particular environmental stresses.

Spectacle Island and the other Boston Harbor Islands have recently come under the auspices of the National Park Service. Visitors to Spectacle Island will have the opportunity to observe the overlay of a reconstructed island with its natural systems set newly into motion.

In the **Stockley Park** project, a landfill near London's Heathrow Airport was relocated to provide a site for development. A primary goal of the project was to blend the site back into its surroundings; in this project technology was employed to hide the site's history. Stockley Park is chiefly a private office park, although it does contain public parks, lakes, and a golf course.

Stockley Park is a highly manufactured site, which has a long history of development and change by humans. The site, a former gravel pit, had been used for the disposal of domestic, commercial and industrial waste to an average 10 m depth. Phased site investigations gave detailed information on the landfill, contaminated groundwater (leachate) and landfill gas.

As shown in Figure 10.7, Arup Environmental developed a reclamation strategy working within the commercial requirements to produce a "clean site" for a business park, and the environmental requirement to reduce pollution of the adjacent canal. Reclamation involved excavation of 2.7 million cubic meters of landfill from the business park and

Figure 10.7 Stockley Park under construction

recompacting it on-site to create a golf course landform. The landfill was then replaced with clean fill materials from the site. Naturally occurring gravels were used to form discrete building platforms that were enclosed by London clay fill dams (also called bunds) supporting roads and lakes within the business park. Costs of the 3.9 million cubic meters of earthworks were minimized by eliminating both the use of imported fill material and disposal of landfill off-site (Arup Environmental).

The business park is completely protected from leachate and landfill gas ingress by perimeter clay dams, keyed into the underlying London clay. Leachate is managed by the combination of a 2.4 km long slurry cut-off wall along the northern site boundary and a drainage system within the site. The cut-off wall prevents groundwater flowing into the site to become contaminated, the water being either deflected around the site or collected and piped directly to the canal. Leachate within the site is maintained below canal level by drains outside of the perimeter clay bunds. Methane vents adjacent to the perimeter clay bunds prevent horizontal gas migration into the business park.

England has a long history of urban settlement compared with the United States, and virtually the whole country has been reworked many times by human intervention. The decision to develop a business park by moving a landfill shows the impact of high real estate values and the increasing necessity to know a site's history (its potential contaminants, etc.) in order to develop it into whatever land use is desired. This remaking of a site highlights the fact that all sites should be viewed as malleable. Stockley Park's location near Heathrow was so valuable that the first building in the business park was completed just 12 months after commencement of the earthworks.

Reclamation: esthetics and economics

Geraldton Mine Project, Geraldton, Ontario

In the **Geraldton Mine Project**, mining tailings are being reshaped for both esthetic and economic reasons. Geraldton, Ontario, (located approximately 200 miles northeast of Thunder Bay) is the site of a closed gold mine. When the mine was operational, the town was booming, swelling to a population of 7,500 in the 1950s. When the mine closed in 1972, citizens either moved away or had to seek employment in the few avenues left. Only 3,500 people live in Geraldton now, but it still serves as a focus for the region. It is home to a regional high school and provides governmental, institutional and commercial services.

Approximately 100,000 highway tourists pass Geraldton on Highway 11 (the Trans-Canadian Highway) every summer, and the immediate goal of this project is to provide this large number of travelers with reasons to stop, stay, tell people about Geraldton, and return. Geraldton is close enough to the United States border to attract American tourists.

The underground gold mine is located 1.5 miles south of the town along Route 11 as one approaches the intersection of Route 584, the major access road to Geraldton. Here, 14 million tons of tailings from the mines have been left. These tailings cover a 170-acre area of land, 27 feet deep. This huge (but flat) pile of tailings greets visitors at the main entrance to the town. Any improvements to this intersection must accommodate future mining activities. This is a very real possibility when the price of gold rises to make the effort economically viable. (Closing and reopening a gold mine is typical of the industry.) Any additional tailings that may be produced must be accommodated in the vicinity of the intersection.

In order to spur economic redevelopment, the town has made the decision to make something (literally) of the sea of tailings. The landscape architectural firm of Martha Schwartz, Inc. developed proposals to improve the tailings' appearance and heighten the visitors' experience of them. Both design alternatives shown in the "Manufactured Sites" exhibition sculpt the flat pile into compelling landforms earthworks that serve as a dynamic

Figure 10.8 "Gold Bar" design alternative

roadway edge and a gateway to the town. The two alternatives: the "Gold Bar," as shown in Figure 10.8, and the "Golden Scroll," as shown in Figure 10.9, were presented at a public meeting held in Geraldton in April 1998. The "Golden Scroll" alternative was selected and it is currently under construction. The landform is designed to be much more than just a powerful visual feature. Trails invite one to walk and explore. Extending the adjacent golf course from nine holes to eighteen onto the tailings is planned.

The new visitor center in the shape of a tower relates to the form of the mine's head frame. While the head frame is closed to visitors, one will be invited to climb the visitor

Figure 10.9 "Golden Scroll" design alternative

144

center and gain a view of the closed mine, the tailings, and the sculptural landform. The mine head frame is of historical significance as it is the only one remaining of the twenty-five or more that existed during the gold mining boom in the area. Tourist information will be provided in the visitor center, encouraging one to consider a meal, an overnight stay, or even investing a business idea in the town of Geraldton. The landform will also have a presence at night as lighting is planned.

Technical constraints were key to the final form of the earthwork. The different types and sizes of earth-moving equipment and their turning radii provided guidelines for the grading plan. A primary objective in the project is to balance cut and fill, and to maintain a maximum total earth moving of 150,000 cubic meters. Cut was kept to a minimum as arsenic levels are higher toward the bottom of the pile. There is a cap at the bottom of the pile of tailings, and there is a maximum of an additional 5 meters that can occur on top. Storm drainage has been maintained and the earthwork will not impede sight lines for traffic safety.

Six to twelve inches of peat topsoil will be added to disturbed areas to aid in revegetation. The soil can support plants, although plants will not be watered. A planting plan for the project focuses primarily on native grasses, especially those golden in color. The master plan also details tree plantings along Highway 11.

The Geraldton Mine project reveals the power of design to remake a wasteland into a new landscape. The tailings are being reshaped into a beautiful and powerful earthwork, although it is much more: this landform is also a cultural artifact, highlighting the location and role of mining in the life of the town. It will also become a site of recreation. The Geraldton project provides an important contrast to the "Effigy Tumuli" project of Michael Heizer. Heizer's project reshaped tailings into earthworks in the shape of various animals that are only recognizable from the air.

The Geraldton earthwork addresses the theme of transformation on many levels: tailings to earthwork, eyesore to amenity, derelict land to recreational resource. The construction of the "Golden Scroll" is a transformative event and embodies the confidence the community has in its future.

Site recovery and public spectacle

The "Green Games," Sydney Olympics 2000, Sydney, Australia
The 2000 Summer Olympics was held in Sydney, Australia, having been designated as the "Green Games" by the Australian Olympic Committee. Besides showcasing the world's greatest athletes, Sydney's Olympics was designed to showcase innovative environmental solutions. Hargreaves Associates, in collaboration with the New South Wales Government Architect Design Directorate (GADD), was commissioned in 1996 by the Olympic Coordination Authority to create a new Master Concept Design for the Public Domain of the site for the 2000 Olympics at Homebush Bay in Sydney, Australia, as shown in Figure 10.10. The result was a project that utilizes emerging technologies in remediation and reclamation, stormwater management, water feature design, urban tree planting, and structural soils.

Olympics have much in common with World's Fairs. Both are seen as opportunities to enhance the visibility of the host city and country, to highlight cultural and technological innovations, and to serve as a meeting ground for the world at the time of the event. Afterwards, they are time capsules of a culture's goals and aspirations for the future. The 1939 World's Fair in New York was one of the largest land "reclamation" projects ever in America, filling wetlands with six million cubic yards of ash and garbage (Hughes 1997: 461). It was not understood at the time that wetlands are crucial to the healthy functioning of a region's ecology. The Sydney project utilized emerging technology to construct new wetlands.

Figure 10.10 Aerial photo of Homebush Bay incorporating the site plan for Sydney Olympics

The Homebush Bay site was reclaimed from a massive post-industrial wasteland of old quarries, saltworks, brickworks, abattoir, and landfill. The Sydney Olympics made a huge commitment to remediating a site with a century of pollution. The environmental innovations were largely invisible to the average visitor, but the emerging technologies created an experience for the viewer that wouldn't be possible otherwise. For example, the dynamic, sculptural forms of the Northern Water Feature's constructed wetlands cleansed storm water, rather than only temporarily storing the run-off in retention ponds as in conventional practice.

While Hargreaves Associates has dramatically revealed the remade nature of the site in the two previously discussed projects in this chapter (Byxbee Park and Crissy Field), it is more subtly revealed in this project. The designers recommended that an existing park, scheduled for demolition within the Olympics site, be incorporated into the Master Plan. This powerful layer of the site's "present" contains mature trees that will provide a shady oasis and a contrast to the architecture. The grid of the old abattoir, that is expressed in the main plaza's paving design, speaks to formal qualities of line and pattern. This heroic plaza was designed to showcase the teeming crowds and, by contrast, their absence. Overall, the project sought to be environmentally conscious, visually arresting, and powerfully memorable.

Figure 10.11 Rendering of Olympic Plaza showing the Northern Water Feature. Fig trees line the Plaza to the left

The Northern Water Feature, as shown in Figure 10.11, connects the urban center of the site with larger landscape systems, including nearby parks and wetlands. Stormwater is collected, passes through a pollutant trap before entering a retention pond, and filters through a constructed wetland. The water is cleansed in the constructed wetland and reused on-site. The wetlands also provide wildlife habitat. The stone and lawn terraces which step down from the plaza to the water's edge, and the pier that extends outward from the plaza to overlook the wetlands, provide a place for visitors to contemplate the role of water on the site.

Environmental engineers and wetland specialists were consulted on the emerging technologies that were key to the built form of the constructed wetland. The requirements of this new technology have two major components: deep pools are used at the beginning and end of the wetland to facilitate the settling of sediments, and a continuous flow of water is employed for filtration. The design of the water feature maximizes water flow by winding back and forth on itself. The forms of the sculpted land and water bodies are intended to express the flow of water through the wetland ponds, while clearly revealing human intervention in the remade site.

Prior to the Northern Water Feature, there was a wetland/retention pond on-site that had failed due to poor planting methods and hard edge treatments. This wetland was leaking and leachates from the adjacent landfill were seeping into the water. Resealing the wetland was necessary, but reforming it was difficult due to not knowing what was under the landfill. Collecting data and mapping the landfill's history was imperative to safely enlarging and resculpting the wetland (Giersbach and March 1998: 5–8).

The Master Concept Design features a long row of fig trees, which line Olympic Plaza. The success of trees within these paved areas depends upon adequate soil volumes and sufficient air and water reaching the root zones. The pavement had to be designed to accommodate immense numbers of pedestrians, while not compacting the soil over the trees' roots. The construction of a continuous trench of soil that is bridged by a concrete deck system that supports paving has emerged as one of the preferred techniques to assure the health of trees in pavement. However, this system is quite expensive, and the possibility of utilizing a structural soil at Olympic Plaza was investigated. Structural soil is a soil mixture that can support pavers and serve as a growing medium.

The Sydney team lab tested soil mixes to design a recipe that could bear high compaction yet provide necessary porosity. A mixture of basalt gravel and sandy loam was selected. The individual pieces of basalt are uniformly cut, and even when the particles are compacted there are still cavities between them where soil, air, and water can flow, allowing space for roots to grow. A layer of pure basalt gravel was laid on top of this mixture (Bennett 1999: 20). Specially designed perforated concrete interlocking pavers facilitate the movement of surface water and air into the root zone. The perforation is a simple modification of the paver that is being used throughout the project.

Sydney's Olympics functioned as both event and image. The "greening" of the 2000 summer games provides a didactic overlay to the Olympics, and discussions of sustainability featured in print and video articles by news crews from around the world as part of the publicity surrounding the spectacle. Highlighting a contemporary attitude to nature via the Olympics is a strategic decision on many levels. The 2000 Summer Games showcased environmental responsibility, marketing, technological advances, and the power of physical design.

Conclusion: the post-industrial landscape and manufactured sites

"Manufactured Sites" reveal a paradigm shift. In the case studies, land is seen not simply as inert matter or a commodity, but as a dynamic web of living systems. The projects demonstrate innovative formal and technological solutions to remediate or maintain the environmental health of the site. A new way of looking at the planning and design of cities is highlighted at the Sydney Olympics: natural processes provide a starting point for urban development, and human use is integrated with these processes. Seeing water, air, soil, vegetation, and habitat as interconnected systems begins to build a city comprehensively in an environmentally responsible way.

This approach to constructing the built landscape has an affinity with the work of Frederrick Law Olmsted, who saw his work on the Back Bay Fens as a sanitary improvement. This project, which created a salt marsh for flood control, was one part of the extensive park system he designed for Boston. The park system created an ecological infrastructure combined with transportation corridors, all while enhancing real estate values. Olmsted's park system accomplished multiple functions in the same space, which is what many of the "Manufactured Sites" accomplish as well.

Robert Smithson's argument that nature and culture cannot be seen as separate is certainly evident in the case studies. Crissy Field in particular blurs these boundaries as it creates a sculptural and functional tidal marsh in the center of a former Army base. Peter Eisenman's theory of the site as a palimpsest is also visible at Crissy Field, where multiple eras of the site's history have been remade. Remediating or enhancing the ecological functioning of the site, and making this process visible to the participant, is a new way of addressing the designed landscape. The goal of many of the "Manufactured Sites" projects is to build landscapes that question dualities of nature/culture and past/present, and to engage those who participate in the site with these questions.

Smithson's proposals to reshape mining tailings would have the effect of making memorials about the history and process of mining (Beardsley 1989: 26). Many of the "Manufactured Sites" projects are conceived in this same spirit; generally the past is not hidden, and the human use and misuse of the site will be legible. These projects, primarily

not nec.
can be both?

in the public realm, are not places of scenery and pleasure. Many of the projects will function as spaces of commemoration: highlighting places despoiled by the mechanical, industrial age and remediated with biological systems in a post-industrial age.

The designed landscape throughout time has been a vehicle to express the human relationship with the earth. The "Manufactured Sites" projects highlight new technologies that work with contemporary scientific data, and the projects also announce that at the beginning of the twenty-first century the interest in expressing the beauty and power of natural phenomena is still very much alive. These projects inspire as well as teach. The "Manufactured Sites" projects are symbolic of a new chapter in the designed environment, one where art and science are highly integrated in the making and remaking of the landscape.

Note

I would like to thank all of the design firms (of the case study projects) that provided information to assist in the writing of this chapter.

Bibliography

Aragon, Louis ([1926] 1994) *Paris Peasant*, Boston: Exact Change.
Beardsley, John (1989) *Earthworks and Beyond. Contemporary Art in the Landscape*, New York, London, Paris: Abbeville Press.
Bennett, Paul (1999) "Innovating in Sydney," *Landscape Architecture* (February), p. 20.
Fisher, Thomas (ed.) (1989) "Eisenman Builds," special issue of *Progressive Architecture* (October), pp. 68–99.
Ghirardo, Diane (1990) "The Grid and the Grain," *Architectural Review* (June), pp. 79–86.
Giersbach, Lisa and Misty March (1998) "Northern Water Feature, Sydney Olympics 2000," unpublished paper.
Groden, Michael and Martin Kreiswirth (eds) (1994) *The Johns Hopkins Guide to Literary Theory & Criticism*, Baltimore and London: The Johns Hopkins University Press.
Hayes, Denis (2000) *The Official Earth Day Guide to Planet Repair*, Washington, DC: Island Press.
Holt, Nancy (ed.) (1979) *The Writings of Robert Smithson*, New York: New York University Press.
Hough, Michael (1995) *Cities and Natural Process*, London and New York: Routledge.
Hughes, Robert (1997) *American Visions. The Epic History of Art in America*, New York: Alfred A. Knopf.
Latz, Peter (1998) Lecture at the "Manufactured Sites" Conference, Harvard University Graduate School of Design, April.
Marton, Deborah (1998) "New Life for the Shoreline," *Landscape Architecture* (October), pp. 82–9.
Matilsky, Barbara C. (1992) *Fragile Ecologies. Contemporary Artists' Interpretations and Solutions*, New York: Rizzoli.
Merivale, John (1978) "Charles-Adolphe Alphand and the Parks of Paris," *Landscape Design* (August), pp. 32–7.
Meyer, Elizabeth K. (1991) "The Public Park as Avante-Garde (Landscape) Architecture: A Comparative Interpretation of Two Parisian Parks, Parc dela Villette (1983–1990) and Parc des Buttes-Chaumont (1864–1867)," *Landscape Journal* 10 (1) (Spring), pp. 16–26.
Nesbitt, Kate (ed.) (1996) *Theorizing a New Agenda for Architecture. An Anthology of Architectural Theory 1965–1995*, New York: Princeton Architectural Press.
Posner, Ellen (1989) "Beyond the Green Quad," *Landscape Architecture* (December), pp. 46–51.
Shapiro, Gary ([1995] 1997) *Earthwards. Robert Smithson and Art after Babel*, Berkeley, Los Angeles, London: University of California Press.
Sutton, S.B. (ed.) ([1971] 1997) *Civilizing American Cities. Writings on City Landscapes by Frederick Law Olmsted*, Cambridge, Mass.: MIT Press.
Vernes, Michel (1984) "Cities & Parks in Opposition," *Architectural Review* (June), pp. 57–61.
Wrede, Stuart and William Howard Adams (ed.) (1991) *Denatured Visions. Landscape and Culture in the Twentieth Century*, New York: The Museum of Modern Art.
Zaiitzevsky, Cynthia ([1982] 1992) *Frederick Law Olmsted and the Boston Park System*, Cambridge, Mass. and London: Harvard University Press.

Chapter 11

Landscape Park Duisburg-Nord:
the metamorphosis of an industrial site

Peter Latz

Introduction

The following essay discusses the possibilities of improvement for the living conditions of a run-down old industrialized region in the northern Ruhr district of Germany, the reanimation of the devastated areas and their regeneration as a landscape for a densely populated territory. It deals with ideas to integrate, shape, develop and interlink the existing patterns that were formed by its previous industrial use, and suggests a new interpretation with a new syntax. The project started in 1991 after an international competition with the masterplan for the whole site and continued with the planning and implementation of many partial projects until 1999. Besides the realization of a 230-hectare landscape park with promenades and footbridges on different levels, with places and gardens and rebuilt water basins and channels, recycling and the dealing with historical contamination were major tasks on this site.

Landscape Park Duisburg-Nord

There, in the so-called Emscher district, the coal, iron and steel industries, shrunken by periodical crises, have left behind a bizarre landscape: spaces torn to pieces and environmental damage, land settlement with swamps and polders, slag heaps instead of floodplain forests and soils polluted with polyaromatic hydrocarbons and heavy metals.

To make this area with 2.5 million inhabitants attractive again for new investment, the fundamental ecological base of the landscape has to be restored. This is the aim of the International Building Exhibition Emscher Park (IBA). The Landscape Park Duisburg-Nord is a key project of the IBA that reflects new ideas about landscape and nature. Old industrial substance is the basis of the park and the existing blast furnaces are its symbol. The metamorphosis of the hard industrial structure into a public park is symbolized by an artifact of seven by seven cast iron plates amidst the blast furnace plant. The original idea for this "Piazza Metallica," as shown in Figure 11.1, was to represent in forty-nine cast iron plates the iron manufacturing process in both its molten and hardened states. Melting and hardening was to be simulated by water flowing and settling on the surfaces. Since the proposed casting would have been very expensive, we found iron plates for the Piazza in the still productive pig-iron casting works. They were used to cover casting molds and had to withstand molten metal at temperatures of 1,600 degrees farenheit. This process caused the plates to take on individual forms of erosion. The plates, each weighing 7–8 tons,

Figure 11.1 Physical nature as a symbolic theme: "Piazza Metallica"

were then mechanically removed from the works and composed in the middle of our selected space. Cleaned of ashes and of casting sediments, they revealed their subtle patterns: ice and iron in a shimmering square during rainfall.

"Physical nature" becomes a symbolic theme. From the first moments of their existence, these plates are eroded by natural physical processes. In this new place, they will continue to rust and erode.

A second theme of the metamorphosis of this landscape is one of "utilization" of the place and the park. Instead of building objects for specific uses, fantasy and playfulness allow the existing abstract structures to function in new ways. Thus our working method is one of adaptation and interpretation, a metamorphosis of industrial structures without destroying them: the blast furnace is not only an old furnace, it is a menacing dragon rising above frightened men, and it is also a mountain top used by climbers, rising above its surroundings. The former ore bunkers become the rock faces of a mountain scenery, or are transformed into enclosed gardens.

The huge ore bunker walls, as shown in Figure 11.2, are used for climbing and the cracks and scars formed by the ore ripping into the surfaces of the concrete allow free-climbers to practice their sport. For the younger mountaineers, we installed an "ascent to the alpine pasture" on a mountain slope with a huge slide, sandboxes and climbing net. Old buildings are used as play-houses or playgrounds. One of the symbols of this area, a pigeon loft, is built into the ruin of the former dormitory for the miners. A diving club has established itself in the deep "caves" where groundwater has formed underground lakes within the redundant ore bunkers. The divers remove impurities and rubbish, and search for adventure in the dark corridors of these subterranean constructions.

The fear of historical contamination has given way to a calm acceptance of the structures, as shown in Figure 11.3. Celebrations are now held each year in the park, and in 1994 at the interim presentation of the IBA, 50,000 people visited the so-called Edward A. Cowper Place. This place was initially controversial because of its location within the blast furnace plant. Nobody believed in celebrating with blossoming trees, "locked up in iron" at the beginning, woven into the framework of the pipes of the windheaters and the blast furnaces.

Figure 11.2 Meta-
morphosis of the
utilization: "climbing
garden"

Figure 11.3 Fear of contamination and acceptance of the structures: "Cowper Place"

In spite of their size, the blast furnace plant with the Cowper Place and the ore bunkers, together with a water channel in the north, and a picturesque harp of rails, only intervene within a small section of the park. This zone includes buildings and architectural structures. As an industrial heritage, they are available now for other purposes: the historic hall just being renovated, the former electrical workshop now serving as a training center, and other buildings that can be used for exhibitions, concerts, or simply for fun and play.

The 230-hectare site largely consists of parts that are left to develop on their own without intensive treatment, or of heavily contaminated parts that have to remain shut off (for example, a coal tar lake). Certain principles were adopted for the park and are revealed in the overall design. There are individual systems operating independently. These layers connect only at certain points through specific visual, functional or merely imaginary linking elements. The uppermost layer is the "railway park" with high-level promenades and the lowest layer is the deep-set "water park." Other individual systems are the connecting promenades at street level and the single fields and clumps of vegetation. All these different layers or single parks are connected by the system of linking elements – either symbolically by gardens or physically by ramps, stairs and terraces. The high dams of the "railway park" cross the landscape wherever they were needed for production. Reaching a height of up to 12 meters, these banks are higher than any of the natural terrain and so offer views of the area that were never possible before. They consist of slags of high pH-value, that are able to immobilize heavy metals. Polyaromatic hydrocarbons exist only occasionally. Therefore the usable surfaces are covered with chips made of dolomitic limestone, also with high pH value.

An area of the park, already worked out by vegetation management is the "rail harp," a bundle of rails in which each rail leads to both a lower and an upper level. It was created by the collective work of engineers, slowly evolving over time as shown in Figure 11.4. Now, a massive land art project has developed, slowly elaborated over again by gardeners. The second system of the "water park" is realized in parts, and large sections are under construction. Natural water bodies no longer exist on the site, due to the extensive sealing

Figure 11.4 Collected work of engineers becoming land art: the "rail harp"

of large areas. The first zone of saturation is contaminated by polyaromatic hydrocarbons. The "Old Emscher," an open wastewater ditch crossing the park from east to west, was carrying non-treated sewage to the river Rhine. For the installation of a new clean-water system we wanted to use the profile of the old construction to avoid contact with the polluted surrounding ground.

The waste water is carried within a 3.5 meter diameter underground main, representing the largest investment in the park and costing more than all the other equipment. It is sealed by a layer of clay which collects run-off from the buildings, hard surfaces, bunkers and cooling basins. Several "water paths" are already working. Rainwater flows in open drains and rivulets and through the existing overhead pipe systems, then falls into the former cooling basins, as shown in Figure 11.5, thus becoming enriched by oxygen. Former settling tanks have been cleaned of 500 tons of arsenic mud, that was squeezed out and loaded into old mines. The tanks are now supplied with clean, clear water, pumped from the depth of the old ore bunkers.

Figure 11.5 New life in the old cooling basins: "new biotope"

Figure 11.6 Wastewater ditch becomes a clear water canal: the "new Emscher"

A wind power plant is installed in the former sintering tower, as shown in Figure 11.6. Water will be pumped from the canal and will fall from several points after a journey through the gardens. This will enrich the biological system with oxygen as well as being an attraction to visitors. Another fascinating layer in the park is the vegetation. Already, in February, a dazzling yellow covers the railway areas, and lichens and mosses grow on the stones of the slag heaps. Steppe-like vegetation grows on the meager soil of coal-soot mixes, as shown in Figure 11.7, on casting sediments and slags of the recently shut down manganese ore depot.

Figure 11.7 Vegetation on contaminated ashes

Some of the ashes are highly contaminated, but could remain if made "not accessible." The preservation of this colorful situation was more important for us than decontamination that would have covered this place completely. Another case is the black waste material of the coal-washing process now colonized by solitary groups of birches. This coal-mine spoil covers a former coke plant that is heavily contaminated with polyaromatic hydrocarbons. There are two alternatives to deal with this problem: layers of clay and the total loss of vegetation, an "eternal grave" of contaminated material, or to have slight gas diffusion over the course of several generations with a corresponding reduction of contamination and limited utilization (such as only cycling, walking, etc.), as we suggested for a project within the former steelworks of Voelklingen (a UNESCO World Cultural Heritage). In Duisburg we preferred the second solution.

The vegetation does not cover the park evenly as it might in "natural" landscapes. Instead, "vegetation fields" lie like single different clumps between the ribbon-like structures of the railway and the water park, covering isolated areas with differentiated forms and colors. They are independent, but within the Ruhr area repeatedly returning vegetation typologies with lots of species (approximately two hundred in Duisburg) from all over the world, called neophytes, are found. These vegetation types made it necessary to train gardeners especially for their management and care.

A largely realized part of the park is the site of the former sintering plant. The transformation of this area represents another form of metamorphosis – that of recycling. Due to heavy contamination the plant had to be almost completely demolished. Only the stump of the smoke stack remained after the blasting. The contaminated material was filled into sealed bags of the sinter bunkers and covered with "roof gardens."

Stones which were not contaminated were ground down to become new soil, new stones, surfaces, and also concrete. The "sintering park" presents itself as a large open space, with the grove of *Ailanthus altissima* growing in ground demolition rubbish, as shown in Figure 11.8, with the ore bunker plant and its secluded gardens and with a

Figure 11.8 Trees in recycled soil: "the grove"

Figure 11.9
Metamorphosis of the
structures: treatment of
contamination, catwalk,
roof-gardens covering the
poison

Roman amphitheater, brightly colored by recycled brick chippings. The stage opens onto seats built to accommodate an audience of 500, and onto the large area which can be used as a festival and concert ground for 5,000 visitors. An old transformer station could be saved and changed into a stage building with colors and some new doors.

Toward the blast furnace, the "sintering place" is framed by a catwalk leading to the "rail harp" on the upper level of the railway park. From this high level walk leading across the bunkers, as shown in Figure 11.9, with a length of 300 meters, the sintering place can be seen on one side, with the blast furnace visible on the other. More importantly, the catwalk gives views down into the gardens. Essential parts of the catwalk are constructed with recycled material. A simple construction, welded on-site, lies on the restored piles of the former overhead railway. Staircases were developed from dismantled maintenance walkways and from old crane bridges. As with the theater and stage house, the catwalk uses bright colors to signal that these parts can be used, in contrast to untouched gray and rusty constructions which remain visibly "off limit." Passing over these zones you can watch the slowly growing bunker gardens, built at various heights and depths within the bunker site. To get into the bunkers we had to cut openings into the massive walls, as thick as 1.20 meters.

Where there were once heaped-up ore and coal deposits there are now flourishing gardens of great variety, as shown in Figure 11.10, which can also be viewed as scenes from the catwalk – for example, a spiral overgrown by ferns and climbers and built with birch wood from the vegetation management of the rail harp. In time, the greenery will dominate the technical constructions of the gateways. So bit by bit another history, another understanding of the contaminated site and of the idea of the "garden," is developing.

In front of the large bunker area adjacent to the blast furnace plant, the contaminated spaces are to be covered, sealed with asphalt, covered with earth once again and planted. I want to explain the principle of this procedure with a description of the project in

Figure 11.10 Ore and coal deposits become the "secret garden"

Voelklingen. An essential part of this site was a factory used for the further development of the liquid and gaseous products of the coke plant. Today, the whole area is contaminated, with polyaromatic hydrocarbons having penetrated the ground and the buildings. These spaces and objects cannot be used, they are far too poisonous for men and animals. The aim was to form a new idea of the elements of nature like earth, water and air, to change them into a new artifact and a technically safe area and to re-utilize it afterwards. Rainwater must never touch these toxic materials, therefore the heaped up demolition rubbish is sealed. A special water system drains the water on the surface. The hill, with its historic contamination deep inside, will be protected by horticultural plants in a grove. Trees are growing in new soil above a sealing layer which will cover the poisonous materials for the next hundred years.

The idea to develop the future out of human destruction has obviously existed for some time. On a journey through the USA I noted a place on the map of southern Oregon marked as absolutely worth visiting: ranges of hills with picturesque cuts and with white, red and orange colors. In the place itself, I learned that these were old mines from the last century where people used water cannons to wash out the mud and, if lucky, gold. For the past 200 years, the ideal image of nature has been a symbolic, transformed and man-made landscape, typified by idealized areas of agricultural production. Such idealizations led to the creation of unique parks, but, as symbols of a past romantic ideal, these landscapes cannot now be restored. These cultural landscapes are as lost to us now as are the social dreams of nineteenth-century Romanticism, and can therefore only fail as ideals for a contemporary landscape. The tasks of dealing with run-down industrial areas and open-cast mines require a new method – one that accepts their physical qualities but also their destroyed nature and topography. This new vision should not be one of "re-cultivation," for this approach negates the qualities that they currently possess and destroys them for a second time. The vision for a new landscape should seek its justification exactly within the existing forms of demolition and exhaustion. We have to ask ourselves which spaces from among the dilapidated and redundant places we want to use and occupy, and which of those have to be changed by the mark of a cultural intervention or the remediation of historical contamination.

Would it not be better to attach to the ideal image of our occidental culture, to "paradise," an oasis in the desert, a place where man has to make his way against the rigors of physical nature? This imagination of an oasis as a garden in desolate spaces is my ideal type of discourse with the nature of old industrial sites, which in their parts can be left to themselves to develop the fantastic images of the future from already existing formations – creating values between art and nature in a way which could never be made by the artist nor mere nature alone.

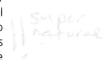

Summary

One question seems to be unresolved: the relation between the function of the park as a municipal open space and its importance for tourism. There are issues arising from it concerning the distribution of funds, the aims and objectives of the new operating company, and of meaning. This becomes apparent with the secondary questions that are discussed vehemently: for example, is there to be fencing and entrance fees to visit the blast furnace plant; what is the level of security for every little part; what is the significance of the children's playgrounds? In reality, each of the individual parts of the park are clearly assigned to defined aims and functions and work out quite well.

Visitors get used especially to the central part where all the events and festivals are happening. Other parts of the park recede into the background. There might also be blame attached to the site of the water park, still under construction, that divides the park lengthways. We shall see how the uses will change once again after its completion in 1999.

We tried to get an answer concerning the behavioral patterns of visitors. A group of students were comparing answers and comments given in the landscape park with those expressed in conventional commercial parks. This attempt at sociological analysis did not result in significant statements – apart from three conclusions:

- The preserved blast furnaces play an important role as a major attraction.
- Visits to the park for individual and personal interest and pursuits are uppermost.
- The primary motive for a visit to the park is the unique atmosphere.

Project information and data

Project

Landscape Park Duisburg-Nord, a project of the International Building Exhibition Emscher Park (IBA).

Location

In the north of Duisburg between the urban districts of Hamborn and Meiderich.

Total area

230 hectares.

Competition

International cooperative competition 1990.
Main planning and realization phase: 1991–9 (end of the IBA), planning and implementation of some minor projects continuing.

Client

Landesentwicklungsgesellschaft Nordrhein–Westfalen.
LEG NRW als Treuhänder der Stadt Duisburg.
(Development Company of Nordrhein–Westfalen in trusteeship for the town of Duisburg),
Gartenamt und Planungsamt der Stadt Duisburg (Authority for Urban Planning and
Authority for Public Green).

Project team

Latz + Partner, Kranzberg/Duisburg.
Latz–Riehl, Kassel, G. Lipkowsky, Oberhausen.
Vegetation concept: J. Dettmar.
Realization by help of numerous citizen groups and associations (IG Nordpark = Community of Interests Nordpark), the Society for Industrial Culture, employment programs for people out of work, workshops with students, pupils and trainees.

Total costs (without buildings and, clean-up of pollution and underground mains)

Thirty million DM, financed by different promotion programs: Oekologieprogramme
Emscher–Lippe, Städtebauförderungsprogramm, Handlungsrahmen Kohlegebiete (Ecological programs Emscher–Lippe, Urban Promotion Program, Skeleton of Action for Coal
Mining Areas).

Further Reading

Beard, P. (1996) "Peter Latz, the Poet of Pollution," *Blue Print*, vol. 7/8 (London).
Broto, C. (1997) "Landscape Parc Duisburg Nord," in *Urbanism – Architectural Design*, Barcelona: Links International.
Cervet, F.A. (1998) "Duisburg-Nord Park," *World of Environmental Design* 11 (recent works by arco editorial, Barcelona).
Diedrich, L. (1999) "A Renewal Concept for a Region, Callwey/München", No politics, no park: the Duisburg-Nord model, *Topos/European Landscape Magazine*, vol. 26 (IBA, March).
Engler, Mira (2000) "The Garden in the Machine, The Duisburg-Nord Landscape Park," *Land Forum – The International Review of Landscape Architecture, Garden Art, Environmental Planning and Urban Design*, no. 5 (Berkeley, Calif.: Palace Press International).
Fabris, L.M.F. (1999) "Emscher Park," *Abitare* 386 (July–August) (Milano: Editrice Abitare Segesta).
Icortesi, I (2000) "Plateau de Kirchberg/Luxembourg, Latz + Partner," *Area* 48 (*rivista di architettura e arti del progetto gennaio/febbraio*) (Milano: Editore Federico Motta).
Holden, Robert (1996) "Hafeninsel, Saarbrücken" und "Landschaftspark Duisburg-Nord," in *International Landscape Design*, London: Calmann + King Ltd.
Holden, Robert (1996) "Making Parks for the Future from the Industrial Past," *Architects' Journal* no. 11 (London).
Lancaster, M (1994) "Hafeninsel, Saarbrücken," in *The New European Landscape*, London: Butterworth.
Latz, Anneliese and Peter Latz (1997) "New Images – the Metamorphosis of Industrial Areas," *Scroope* 9 (Cambridge Architecture Journal).
Latz, Peter (1993) " 'Design' by handling the existing," in V. Uitgeverij Thoth (ed.) *Modern Park Design*, NL-Bussum.
Latz, Peter (1996) "Emscher Park Duisburg," in *Landscape Transformed*, London: Academy Editions Ltd.
Latz, Peter (1999) "Water – a Symbol for the Ecological Rehabilitation," in *UIA Work Programme: Architecture of the Future*, ed. The Japan Institute of Architects, May, Tokyo.
Moore, R. (ed.) (1999) *Vertigo: The Strange New World of the Contemporary City*, London: Lawrence King Publishing.

Morris, S. (1999) "An Industrial Legacy and a New Beginning," *Byggekunst* (The Norwegian Review of Architecture) 81(8) (Oslo: fagpressen).

Parker, A. (1996) "The Beast Breathes Again," *Landscape Design*, no. 254 (October).

Raver, Anne (2000) "Splendor in Rust Belt Ruins," *The New York Times*, February 10.

Sheard, P. (1996) "The Road to Recovery," *Horticulture Week,* vol. 19 (London: Blueprint Media Ltd.).

Wachten, K. (ed.) (1996) *Change without Growth? Sustainable Urban Development for the 21st Century* (VI. Architecture Biennale Venice), Braunschweig/Wiesbaden: Vieweg & Sohn Verlagsgesellschaft mbH.

Weilacher, U. (1996) "Die Syntax der Landschaft, Peter Latz," in *Between Landscape Architecture and Land Art*, Basel/Berlin/Boston: Birkhäuser.

RESPONSE

Terra-toxic

Niall Kirkwood notes in Chapter 1 that remediation of our contaminated sites should not be viewed, as President Clinton had asserted in his 1997 State of the Union address, simply as parks or poison, but, more realistically, as "park and poison." This approach implies the coexistence of poisons within our parks. The question is what technologies and design theories can be employed to achieve a safe coexistence and, in the process, reclaim these degraded sites for productive uses.

In his work at Park Duisburg-Nord, landscape architect Peter Latz brings this construct one step further, and, merging poison *and* park, has created *place*, thus "park" plus "poison" equals "place." Latz, perhaps more than any other landscape architect thus far, has located and investigated opportunities in this equation. Latz is particularly unique in his ability to esthetically read and appreciate the post-industrial landscape and to find symbolic icons, to reconnect our contemporary culture to its working past. His interpretation of the degraded site then informs his design and remediation decisions. His use of technologies to readapt the structures reveals, without disguising, the order, scale, and full character of the site. A precedent for comparison exists in the United States, in Richard Haag's Gasworks Park in Seattle, Washington. But while Gasworks Park pioneered the incorporation of industrial structures in a public park site, Latz's venture extends well beyond preservation. He reinterprets the historical structures for contemporary uses that engage the community beyond simply reading the past. Park Duisburg-Nord offers a new hybrid, a new model, an "industriopark" celebrating community and culture.

The Duisburg-Nord Park project offers a holistic approach to re-manufacturing a contaminated industrial site into a community park. This case study offers a clear example of how historical preservation and remediation technologies can be joined to establish our cultural connections to our nineteenth-century industrial heritage, without sentimentality and with a deep respect for the integrity of the existing forms and structures. As Latz describes, the "Old industrial substance is the basis of the park." Examples of remediation projects that come wholly into public hands are uncommon. The typical model for site remediation may include open space set aside, but is developed through the private sphere, with the constraints of promoting an unassailable corporate image, as well as profit margins and liability. Duisburg-Nord offers a unique model since the intent has always been to use the entire site as a community park facility.

To incorporate the industrial structures into the working park, Latz reinterpreted these elements, asking how they might provide a utilitarian amenity. The result is a provocative synthesis of the physical and symbolic histories of the site, forming an armature for engagement. Existing industrial structures become climbing walls, buildings are transformed into play-houses, and former bunkers, now under water, become "caves" for local diving enthusiasts. The design of Duisburg-Nord, was, and still is, unique for its celebration of these industrial remnants as cultural icons and for the use of these remnants for community recreation and enjoyment. The site, now being used for celebrations and civic events, has been re-manufactured, the restored buildings now function as training centers, exhibition halls and "play-scapes."

As he did with the existing structures, Latz, in his approach to the site work, responds to the patterns of ecological disruptions, allowing these patterns to determine the shape of a new evolving ecology for the site. The decision to leave contaminated soils on-site offered an opportunity to reveal the patterns of the disturbed vegetation, thus mapping the contamination. The imprints of the spoils, buildings, infrastructure and soil contamination become design determinants, providing an underlying ordering structure. As Latz remarks, "The vegetation does not cover the park evenly as it might in 'natural' landscapes. Instead, 'vegetation fields' lie like single different clumps between ribbon-like structures of the railway and the water park, covering isolated areas with differentiated forms and colors." The vegetation responds to the contaminants, much as it would to differing soil conditions. The decision not to seal or cap parts of the site, accepting slight levels of gas diffusion instead, and allowing the healing to occur over a longer period of time, created an opportunity for a differing esthetic to emerge from the historic and contemporary processes on-site. In situations where toxicity is high and human access is restricted, soils have been moved into existing structures, covered with a growing medium and planted, transforming these existing bins into a series of viewing gardens, seen from the catwalks above. Some are temporal gardens, revealing decay, and as they break down add nutrients into the soil. Others are planted with strange plants, chosen or establishing themselves as species tolerant of the contaminants. A color-coding system takes the place of signage, indicating what areas are accessible and which are restricted access.

As with Walker and Owen (see Chapter 8), Latz utilizes the restoration of ecological processes as a means for development – in this case for community use. To achieve this, relatively simple technologies were implemented to reintroduce ecology back into the site. In using technology borrowed from the past, wind power pumps clean water from a nearby canal, channeling it through a series of gardens and enriching the biological system with oxygen. The use of "green" technologies to heal a nineteenth-century industrial site is itself a strong metaphor. As Latz reveals natural process through the reintroduction of water, he also utilizes "physical nature" as a symbolic theme. This idea, the marking of natural processes, is repeated throughout the park. By introducing cast steel plates into the plaza, Latz references former industrial processes. The plates serve as barometers of erosion and entropy, revealing the natural processes through the rusting and corrosion of the material. The actual and symbolic reference reflects Latz's belief that time will transform the site, that the process is a continuing one, evolving as site conditions change, and that the designer is simply a facilitator. Through these overlapping histories, within a burdened site "the 'garden' is developing."

A number of questions are raised in Latz's approach, suggesting many avenues of research now that this and other projects are completed. Part of this research must extend beyond the designer's intent, engaging the users and public, assessing their perceptions and responses to the work. For example, is Park Duisburg-Nord viewed by the designer and/or users as a step toward greater environmental health, or does Latz step aside, removing himself from judgment values of good or bad health, instead focusing on natural systems of change?

Clearly there is much to learn and admire in Park Duisburg-Nord. Latz's approach, one of re-adaption, preservation and on-site treatment, presents a new paradigm for a "park." As a model, it offers an option to the demolition and encapsulation of the site, one that can and should accept "their physical qualities, [and] also their destroyed nature and topography." As Latz advocates, we should not borrow an esthetic from the romantic era of beauty parks, nor the model of park used solely for recreation, but should respond to the forms inherent in our industrial history and processes. He sees a symbiotic relationship where the parts "can be left to themselves to develop the fantastic images of the future from already existing formations – creating values between art and nature." Here again questions arise, questions of legibility and narrative. Can the design for Park Duisburg-Nord be read as a narrative, a story telling the past, present and future of the site? If this is the intent how is it being read and what is the response?

Of the many benefits offered by this project, maybe the most obvious is the strong esthetic, and the unique form of place that emerges when the design is informed by the history of the site. This clearly is a call for designers to approach a site from a holistic vantage point, one that sees the layering of benefits as a part of site remediation. It is clearly an approach that challenges the skills of designers and their ability to establish a meaningful design, provide a community benefit, and redress the toxic nature of the site.

As a model of reuse, it has the added benefit of less traffic and hauling from the site, resulting in reduced negative impacts on the surrounding communities. By minimizing demolition, the potential for dust-borne toxins is reduced, and demolition costs saved. In using existing structures such as the containment bins, excavation costs have been saved. Is this a model that can be used at other sites? There were several unique opportunities existing in this site that may not be found in other situations; however, many of the strategies offered in this project may be incorporated, in part, at other sites.

To accomplish Duisburg-Nord, an engaged community providing strong support was critical to the success of the project. A population density within close proximity to the site is necessary to justify recreational programmatic elements. Access to and maintenance of the site and organized activity planning are also critical components to ensure year-round use. The same level of support from the municipality, state and counties is critical in both the developmental and the post-construction stages.

Is this model possible in the United States? As McNeil and Lange points out in Chapter 6, "Environmental regulations and liability are generally regarded as the single most significant deterrent to site development." And in a project such as Duisburg-Nord, the long time-frame, the on-site gas diffusion and the focus of people engaged with the land all raise legal and liability issues that, while not insurmountable, would have to be negotiated through local codes. However, with more recent legislation geared toward setting clean-up criteria to be consistent with intended use, some latitude may exist to allow these alternate uses to be incorporated in future remediation projects in this country.

In reviewing Park Duisburg-Nord, questions of cultural differences arise. Latz states that in the end the "fear of historical contamination has given way to a calm acceptance of the structures." What was the nature of public trust in the authorities to ensure that a healthy habitat for recreating was in fact possible? How did Latz and/or the authorities gain this public trust and is the site periodically monitored to alleviate any current or future public fears? Can design itself really overcome public paranoia or is outreach and public education an essential component of this type of remediation? Are there differences between the German and the American public in their reliance and trust in the public authorities? Many Americans have prevailing negative associations with clean-up sites from Love Canal in the 1960s to the Hanford nuclear site which continues. After gathering new data that indicates the environmental health risks at Gasworks Park have not been properly resolved the site is being re-evaluated. One of the most pervasive issues stemming from this process are the difficulties in finding public consensus on how to address the problem, and, for the public, which of the varying scientific authorities to align with. The re-examination of the safety issues, particularly soil contamination and safe play, has spawned many contentious public meetings, indicating that a cloud of continuing reassurance hangs over the concept of site remediaition for public parks.

A project such as Duisburg-Nord may not be possible in a privately funded project where the return on investment is calculated in currency, and housing, office space and retail are critical to the economics of a project. However, this model may present some options where adaptive reuse of the structures, revealing of the natural and technological processes, and the utilization of the site for recreational use may provide a greater return at less cost.

Finally, does this project offer some technologies that should or can be adopted in other projects?

Park Duisburg-Nord provides a bridge, one that borrows from earlier attempts at preservation, utilizes remediation technologies, and offers an inclusive process that brings community members, public sector administrators and a diversity of professionals to the same table, where each is heard and their values and ideas incorporated into the project. This project, from a design and community development point of view, offers a challenging model for further exploration and research in the re-manufacturing of damaged sites.

Chapter 12

Science, engineering, and the art of restoration:

two case studies in wetland construction

Wendi Goldsmith

Introduction

Landscape architects are faced with many opportunities to integrate esthetic design with human use, enjoyment, and education as well as stewardship of the natural environment. Virtually every site under development must also address issues related to stormwater management, whether by legal mandate or simply for practical reasons. Management of water on-site can be handled with conventional, ugly, and single-minded treatments exemplified by the traditional rock-lined detention basin with a chain-link fence around it, or through attractive landscape water features which invite contemplation while promoting wildlife habitat and water quality.

Past and current land-use activities typically contribute to degradation of the qualities of a natural resource by displacing native plants and site-adapted soils, upsetting the hydrologic balance of the watershed, fragmenting and disconnecting remaining adjacent habitat features, and generally introducing contamination in forms ranging from excess nutrients to the dumping of industrial waste. A responsible land ethic requires stewardship that includes restoration of past damage while avoiding continuing degradation. The legal costs of addressing wetland, stream and stormwater issues already form a considerable burden to public and private interests. By working in collaboration with natural scientists, resource managers, and engineers, today's landscape architects can design projects that restore natural ecological functions in degraded riparian areas.

In this essay, two case studies are presented and discussed as examples of landscape design that integrates resource restoration with site planning. The first, located at the Federal Medical Center in Devens, Massachusetts, shows the potential for biological enhancement of a utilitarian stormwater system. The second, a wetland restoration in Vinalhaven, Maine, describes the potential for re-creating severely altered tidal marshland. This second project represents an innovative collaboration between The Bioengineering Group of Salem, Massachusetts and Aviva Rahmani, an environmental artist.

1 Federal Medical Center, Devens, Massachusetts

A bioengineered pond and wetland system, located on the site of a decommissioned army base at the Bureau of Prisons Medical Center in Devens, successfully combines stormwater quality and discharge control with ecological enhancement. The project created three interconnected, heavily vegetated basins along an existing stream course to improve water quality, stabilize soils, and provide wildlife habitat. It incorporated specialized bioengineered, vegetated materials to establish vegetation reliably and quickly allow the system to regulate stormwater flow for events up to the 100-year storm.

Designs for this site were affected by topography, security considerations, and concerns for the preservation of sensitive natural resource areas, including a National Wildlife Refuge and a downstream lake. The most cost-effective site layout threatened to affect an already altered and disturbed stream and draw resistance from regulatory agencies. An elaborate and expensive stormwater system was originally planned on higher ground, away from the stream, but project designers eventually decided to construct the stormwater system in the existing stream course. Although altering an existing resource area is controversial, they realized that they could create an ecologically beneficial and esthetically pleasing system that would also serve the engineering function of discharge control and pollutant removal.

This project's combination of many features in a multi-purpose channel and basin complex garnered the 1997 Environmental Achievement Award from the International Erosion Control Association. The award recognized the successful completion of a major construction project adjacent to a sensitive wildlife habitat area, and its ability to withstand an extreme storm in its first season without the release of damaging discharges of sediment to downstream resources.

Project participants

One notable aspect of the project was the coordination necessary to implement the innovative and somewhat controversial approach. Carol R. Johnson Associates, Inc., of Cambridge, Massachusetts, a large landscape architecture and environmental planning firm, served as site team design leader. The Bioengineering Group, Inc., a design firm with a small staff of engineers and natural scientists specializing in water's edge stabilization, restoration, and revegetation, addressed technical issues related to vegetative stabilization and water quality improvement. Stubbins Associates, Inc., an architectural firm based in Cambridge, consulted on project implementation; Haley and Aldrich, Inc. of Charlestown provided geotechnical engineering support; and Bryant Associates, Inc. of Boston offered civil engineering expertise. Specialty bioengineering products such as prevegetated fiber mats and coconut fiber rolls were supplied by Bestmann Green Systems, Inc., based in Beverly, Massachusetts. With support from the other firms, The Bioengineering Group, Inc. prepared a full set of plans, cost estimates and specifications, including detailed quality assurance guidelines addressing contractor qualifications, approval of submittals, and a work sequence program. Ongoing communication with the contractor, Emanouil Brothers, Inc., of Chelmsford, and regular site inspection visits ensured that work progressed smoothly.

Description of the stormwater system

The stormwater system consists of three interconnected basins located within the existing stream course. In addition to providing water quality control for stormwater flows from typical storms, this newly constructed basin/channel system regulates stormwater discharge for extreme events up to the 100-year storm.

The stream enters the system from a pre-existing culvert in a headwall. From this point, carrying run-off from a watershed of 21.7 hectares (55 acres) and a hillside spring, it enters the first basin. Sized at 0.05 hectares (0.13 acres) and designed with a permanent wet pool to trap 50–70 percent of incoming sediment, this first basin provides primary treatment. From its edge, the stream flows to Basin 2, a 0.10 hectares (0.25 acres), heavily vegetated wetland which provides water quality improvement via filtration, biological treatment, and further sedimentation. Soils, fine grading, and plant materials were carefully specified for this pond to maximize fine particle removal, nutrient uptake and hydrocarbon breakdown. Baffles built of natural materials increase surface area for biologic treatment and direct flows to reduce short-circuiting, which would decrease residence time.

The stream discharges from Basin 2 over a constructed rocky 0.5 meter (1.64 feet) cascade before entering the third and largest basin, 0.30 hectares (0.75 acres) in size. The cascade induces mechanical aeration to increase dissolved oxygen, and provides a visual landscape feature that people can enjoy from benches on a nearby overlook. In addition, the third basin includes a wildlife habitat island and a wetland border consisting of many attractive native grasses, sedges, rushes, wildflowers, shrubs, and trees. The size and depth of the third pond supports a diverse aquatic community, while native edge vegetation offers a variety of terrestrial wildlife habitats. With the substantial improvement in water quality provided by the first two basins, the third basin provides important esthetic functions. It also further dilutes and cools pavement-warmed rainwater and allows expanded residence time for passive treatment of water chemical parameters. By the time the water reaches the earth berm with a notched concrete weir at the end of the system, it has been cleaned and controlled to preserve the natural channel downstream.

The landscape adjoining the basin was preserved intact or regraded with naturalistic landforms. A selection of native trees, shrubs, and meadow grasses enhances the setting. This planting also maximizes the interception and infiltration of stormwater, in contrast to simple turf, which yields relatively high run-off and excess nutrients due to fertilization. The chosen species offer food and habitat as well as resting and nesting sites for a variety of birds and small to medium-sized mammals. As a result, fox are frequently observed preying upon rodents, and a troublesome goose population has decreased to a fraction of its former size due to fear of predators.

Bioengineered vegetative installations

The entire detention system is ringed by bioengineering treatments to provide immediate and long-term stabilization to the watercourse's nearly 600 meters (2,000 feet) of bank. The treatments around each basin and along each connecting channel were selected based on water level and flow conditions.

In stress areas of high flow velocity or steep banks, 0.3 meter diameter coir (coconut-husk) fiber rolls were placed approximately at normal water level to protect the banks as shown in Figure 12.1. After placement, the fiber rolls were planted with a varied selection

Figure 12.1 Placement of coir fascines for pond edge stabilization

Figure 12.2 Pre-vegetated coir modules allow rapid establishment of erosion control functions

of vegetation plugs, including sedges, rushes and bulrushes. Lesser quantities of flowering natives such as blue-flag iris (*Iris versicolor*), cardinal flower (*Lobelia cardinalis*), and monkey flower (*Mimulus ringens*) were also planted to enhance esthetics and attract pollinators.

In lower stress areas, pre-vegetated coir mats were installed to establish healthy, robust vegetation quickly. Single species pallets of iris, sweet flag (*Acorus calamus*) and other species provide an esthetically pleasing, verdant coverage on the lower banks. Mixed species carpets such as these provide similar benefits and increased species diversification.

The wetland at Basin 2 was constructed around and within a sinuous, highly meandering channel, using a combination of bioengineering materials. This wetland was developed using pre-vegetated coir mats as shown in Figure 12.2, containing a pre-grown mixture of sedges, iris, Canada manna-grass (*Glyceria canadensis*), rice cutgrass (*Leersia oryzoides*) and blue-joint reedgrass (*Calamagrostis canadensis*). All of these plants are tall, clump- or sod-forming species that are highly resistant to flow

Basin 2 was shaped around three tight meanders created by bioengineered baffles, extensions of plant mats, which cover from one-half to two-thirds of the basin width. The mats were pre-vegetated with soft-stem bulrush (*Scirpus validus*), a tall, flow-resistant, and prolific native wetland plant. The area located between the vegetated baffles was planted with large plugs rooted in coir, clustered by species based upon stem strength for impact resistance and esthetic effect. Density of planting also varied, with pockets of open water to maximize wildlife habitat opportunities.

On the steepest slopes of the berm flanking the weir in Basin 3, where the impact of high flows and water level fluctuation is predicted to be greatest, live brush layers were installed over a fiber roll toe for stabilization. Cuttings from live wetland shrubs were placed in eight rows, each separated by a layer of topsoil. The design called for a gradation in species placement from bottom to top based upon inundation tolerance. The lowest row, placed just above normal water height, contains the most water-loving species, such as red osier dogwood (*Cornus stolonifera*) and buttonbush (*Cephalanthus occidentalis*), while the uppermost rows contain more drought-tolerant species, such as arrowwood viburnum (*Viburnum dentatum*), willows (*Salix* spp.) and elderberry (*Sambucus canadensis*). Due to the species used, as well as the design and use of the berm, these plantings pose no structural risk.

Construction phase

A comprehensive erosion and sediment control plan based on careful sequencing of the work ensured the project's success. Seeding and mulching took place throughout the construction process to minimize exposed soils. Excavation of the new basins was completed prior to major site regrading to provide construction-phase sediment traps. Bioengineered erosion control measures were installed around the new basin and channel edges as they filled with water. Special fencing consisting of criss-crossed twine was erected to protect new wetland plantings from goose herbivory. The project was installed in phases, with Basins 1 and 2 completed in late June 1996 during a two-month dry spell, and Basin 3 in August 1996. Basin 2 required daily irrigation during the weeks immediately following installation due to dry weather. The final earthworks and plantings associated with the berm and weir were completed in early November 1996.

Post-construction conditions

Soon after planting in July 1996, New England was visited by the remnants of a hurricane that produced over four inches of rain within a twenty-four hour period in parts of central and eastern Massachusetts. Nearby streams were reported to have reached between the one- to two-year flood stage. This storm represented a severe test to the newly installed treatments that had only a few unseasonably dry weeks to begin establishment. However, when the floodwaters receded after the hurricane, no damage was visible in the system. In addition, despite earlier concerns about the ability of Basin 3 to hold water, it filled to approximately the desired level.

Water in the basins has remained at desired levels as the system has evolved. Wetland plantings have also remained intact, and have in fact flourished. Basin 2, the constructed wetland, performed particularly well; its new plantings greatly increased their biomass within two months. Also encouraging was the vigorous show of flowers by many species such as the water plantain (*Alisma plantago-aquatica*). No erosion problems were evident by late summer, and later site visits have confirmed that the vegetation at the site is performing quite well as shown in Figure 12.3.

Figure 12.3 View across the pond showing wildflower meadow, tree and shrub planting, wetland edge and concrete weir

170

Closure

The successful design and installation of this stormwater management system was possible only through the collaboration and partnership of the design team and, ultimately, the contractor. Design team members recognized the opportunity for integrating water quality control and flow detention with esthetic and ecological benefits, and were willing to work to implement this innovative approach.

The result is a highly functional system from an engineering perspective, with lower maintenance costs than conventional facilities and plantings. The site is also greatly enhanced by natural beauty that changes with the seasons; the sight and sound of flowing water and wildlife, visible evidence of the protection of water quality during and after construction, demonstrate the project's effectiveness.

2 Roberts Harbor tidal marsh and riparian zone restoration, Vinalhaven, Maine

In 1990 Aviva Rahmani, an environmental artist, purchased a parcel of a former quarry at Roberts Harbor on the island of Vinalhaven, Maine, for the purpose of reclamation. Once buried under tons of quarried granite, the site had long since lost its cover of native vegetation and undergone dramatic changes in soil structure and hydrological conditions. Ms Rahmani took it upon herself to restore the drainage patterns, improve soil fertility, and plant communities that had once thrived there.

After several years of aggressive compost application and transplantation of native seedlings and garden plants, the barren rock-strewn uplands adjacent to the shoreline marsh showed signs of new life. Ms Rahmani then turned her attention to the restoration of an intermittent stream channel, buried under hundreds of tons of quarry waste, which drains to the marsh. She drew inspiration from a pocket of native vegetation that had survived despite the extensive damage. Located between upland ecosystems and coastal tidal marshland, this channel plays an important ecological role. It creates linkage for a Class A migratory bird fly zone, and ensures the health of local ecosystems by controlling water transport.

The Bioengineering Group has been an ongoing partner for this project since 1994. Responsible for project design, consulting, and engineering, it has used state-of-the-art bioengineering techniques, including placement of coir rolls and fabric, to stabilize plantings of native grasses and other vegetation to re-establish natural plant communities.

Introduction and site description

Despite its apparent isolation, Vinalhaven Island played a role in the development of North America, and shows the scars of industrial impact. Rock from Oatmeal Quarry, lying adjacent to the marsh, was used in the construction and architectural detail of many notable buildings and structures, including the Cathedral of Saint John the Divine in New York, the Saint Louis Post Office, and piers for the Brooklyn Bridge. During the peak years of excavation, the marsh site was buried under thousands of tons of granite blocks, small granite chips from shaping, coal and slag from "burning out" the rock, and even rails from the freight car system used to haul quarried rock to the stone pier that still juts into the harbor. It had also been used in recent years as the town dump.

Research showed that salt marsh conditions were present at Roberts Harbor prior to disturbances caused by extensive quarrying. Numerous clumps of salt marsh vegetation existed on the 0.1 hectare (0.25 acre) site, literally peeking out from under granite blocks, including smooth cordgrass (*Spartina alterniflora*), salt meadow grass (*Spartina patens*) and black grass (*Juncus gerardii*). The presence of these species indicated that the proposed project was a restoration of natural conditions, and that the area could support the reintroduction of wetland plants. In addition, the feasibility of the project was supported by successful restoration efforts by Ms Rahmani on adjacent riparian land. Prior to the start of work on the channel and marsh, six years of restoration on one neighboring hectare (2.5 acres) had restored the free flow of water from surrounding uplands disconnected for over a century by granite excavation and fill. This companion project laid a critical foundation for the re-creation of the wetland by re-establishing natural hydrologic conditions.

With the realignment of economic needs and societal perceptions restoration of the lost salt marsh has become a valued activity. The wetland restoration has provided critical habitat for wildlife, including migrating waterfowl, and created esthetic and ecological improvements that provide important economic benefits to the community. It has also created a model for future restoration of natural functions to other man-made structures along the Gulf of Maine.

Bioengineered vegetative installations

The Bioengineering Group's experience with salt marsh restoration projects provided the basis for a reliable and site-specific design for Roberts Harbor. The use of bioengineering improves upon earlier marsh restoration techniques in its use of coir (coconut) fiber-based, biodegradable planting modules as shown in Figure 12.4.

This technique leads to maximized stability in the tidal environment, which is particularly important here due to the amplification of wave action by the tapering shape of the harbor basin. The pre-vegetated coir fiber modules also afford the substantial advantage of anchoring into the substrate, including a rock outcrop, which is impossible with more common planting methods. Further, the coir medium encourages dense plant rooting, and both the coir and roots eventually intertwine with the soil below. As flushing of the tides mobilizes sand and other sediments, these are deposited within the coir material, ultimately producing a soil substrate as the coir decays.

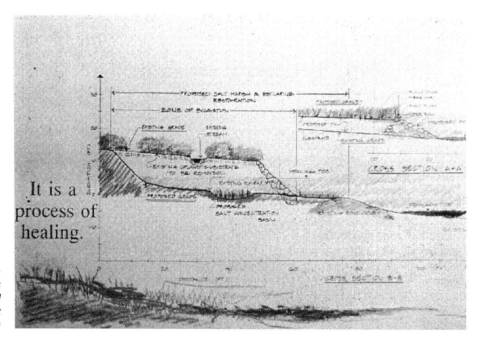

Figure 12.4 Design cross-section and artist's rendering of proposed fill removal and salt marsh restoration

Figure 12.5
Construction view of
tidal inlet regrading

Construction phase

In contrast to restoration efforts that emphasize transformation and improvement of vast acreage, this project focuses on a thorough restoration for a small area; its motto has become "wetlands and other natural resources were lost by increments, and must be restored by increments."

The completed project restored approximately 240 square meters (2,600 square feet) of man-made land to salt marsh conditions through riprap/fill excavation and revegetation as shown in Figure 12.5.

Some 138 cubic meters (180 cubic yards) of quarry waste were excavated from the site and removed from the premises. Following rock and debris removal, a variety of revegetation methods were used, including the placement of pre-vegetated fiber rolls (biodegradable coir fiber cylinders), and planting of individual coir-wrapped plugs. Where bedrock was present at the surface, cables encircling the rolls were bolted directly into the rock itself. The periodically inundated zones of the marsh were thickly mulched with cordgrass hay held down by a woven coir mat pinned in place with existing rocks and wood stakes, using a high tensile strength geogrid material to further anchor the plantings when necessary. Individual plant plugs were placed through the fabric and secured between rocks in the substrate's interstices. The entire project was constructed in March 1997 to allow a full growing season for root development prior to the stormy New England autumn and even more problematic spring thaws.

In terms of configuration, the outermost components of the design were the pre-vegetated fiber roll, planted with smooth cordgrass (*Spartina alternifolia*) to provide a strong frontal barrier against tidal impacts. The fiber roll was placed at the mean high tide elevation to maximize root contact with infiltrating tidal flow, while protecting them from wave action. Immediately behind, cordgrass was placed upon a slight embankment. Beyond the embankment much of the remaining area was planted with a diverse mixture of low–high salt marsh and fringe species as shown in Figure 12.6, including salt meadow grass (*Spartina patens*), spike grass (*Distichlis spicata*), black grass (*Juncus gerardii*), and switchgrass (*Panicum virgatum*)

A 200 square foot salt concentration basin was located toward the restoration area's landward side and left unplanted in order to concentrate salt by the evaporation of trapped seawater. This saline "panne" area provides a distinct boundary with the freshwater zones upstream, and serves to deter opportunistic species such as common reed (*Phragmites australis*).

All soils encountered during the excavation process were salvaged and applied on the surface of the rocky shoreline above tidal elevations, then heavily seeded and mulched.

Figure 12.6 View down tidal inlet after the first growing season

Various shrubs and even small trees growing on the excavated area were gently bucketed out and replaced at final grade with the aid of a skilled brackhoe operator.

Closure

Ms Rahmani has worked both independently and in collaboration with other artists to create a series of artforms related to the restoration effort, including oil paintings as shown in Figure 12.7, watercolors, stained glass, performance pieces, videos, on-site granite sculpture, and a website.

Figure 12.7 Oil painting, as well as stained glass, performance art, video and marsh restoration, was part of the artists' multimedia work

174

She is currently artist-in-residence at large at the Wells National Estuarine Research Reserve, in Wells, Maine, where she works in collaboration with Dr Michele Dionne, Chief of Research. Dionne and Rahmani are experimenting with linkage theory models that incorporate hard science with an artist's observations of visual patterns. A film is in development from that work.

The site itself is designed to draw people in with colorful, appealing plants, then teach more as people stay and study the complex features there. Ms Rahmani has planted annuals and perennials around the restored salt marsh to promote visual interest and integrate the feature with the adjoining upland gardens. To the casual observer, the site is green and alive, in striking contrast to the sterile fill that forms the adjacent pier. Ms Rahmani's studio is only steps away from the new marsh, and she continues to generate varied artistic works beyond the marsh itself. The visual monitoring of the site is periodically posted on Rahmani's website (www.ghostnets.com).

The Roberts Harbor salt marsh restoration offers an excellent opportunity to demonstrate the superior results obtained through the use of the bioengineering approach. Following several storm events, soil and plants have been refurbished and nurtured as needed to support the new marsh until its full critical level of maturity has been reached. Results have been very encouraging. Though it is much easier to destroy than rebuild a coastally exposed marsh, and any efforts to do so must correctly account for the highly dynamic physical and biological processes that affect the site, this project has demonstrated the feasibility of such projects on even the most damaged sites. The application of new bioengineering techniques for soil stabilization and revegetation represents a powerful new tool for the management of coastal wetlands, and art is likewise effective at making people aware of the need.

RESPONSE

More bio than engineering?

The description that follows lays out the underpinnings of the work carried out by Wendi Goldsmith and her colleagues in bioengineering and landscape restoration design.

"Bioengineering" is an approach to erosion control, bank stabilization, wetland restoration, water quality improvement, and habitat enhancement, which uses living plants, and organic structural elements and materials as basic construction modules. The term "bioengineering" describes several methods of establishing vegetative cover by embedding a combination of live, dormant and/or decaying plant materials into banks and shorelines. Variations of the term are also in use, including "biotechnical slope protection," "biogeotechnical erosion control," "soil bioengineering" and "hydro-bioengineering"; these terms sometimes further differentiate sub-fields of bioengineering or differentiate it from the unrelated fields of biomedical technology and genetic engineering.

Bioengineering systems work together with the dynamic processes of erosion and sedimentation along the bank or shoreline to provide ongoing stability, starting at the time of installation. These systems generally employ mechanical components made of biodegradable material, sometimes reinforced with synthetic materials. The use of "hard" barriers such as rock is held to a practical minimum, with the design purpose being to establish a vigorous, sustainable native or naturalized plant community as the primary erosion control mechanism. Bioengineering provides all the benefits of vegetation more quickly and reliably than conventional planting techniques. These benefits include bank stabilization, habitat enhancement (food and cover sources and temperature control for aquatic and terrestrial animals), purification and filtration of overland run-off, contribution of organic matter (nourishment) to land and adjacent water bodies, and esthetic improvement. A great strength of bioengineering is its flexibility. Bioengineering methods can be used separately or in widely varying combinations to fit the great diversity of conditions to be addressed. Combined with sound application of principles of lake management or fluvial geomorphology, bioengineering can serve as a valuable technology in lake and stream restoration.[1]

Of interest to those who address the issues of restoring manufactured sites is the ability to use a broader series of engineering and design tools, with biodegradable site materials being applied to both clean up contamination (phytoremediation) as well as repair or restore damaged earthworks, slopes and water systems (soft engineering). The significance of integrating bioengineering as opposed to conventional environmental engineering techniques is demonstrated in the two case studies, Devens, Massachusetts and the wetland restoration in Vinalhaven, Maine. The sea-edge project carried out in conjunction with environmental artist Aviva Rahmani is possibly unique in terms of its geographical location and the impact of the restoration on the "larger physical and biological processes." However, as Goldsmith explains, the project hinged on "the restoration of natural conditions." Questions that arise relate to the restoration of salt marsh plantings and stabilization techniques and the ability for small project areas such as this to serve as larger bio-models for regenerative form. What do the issues of scale bring to these "incremental" models; can the same benefits be brought to larger sites with an economy of coverage? The case study also demonstrates the ability to include other participants into the regeneration of manufactured sites. With the exception of artist Mierle Ukeles, most closely associated with the Fresh Kills landfill discussed in Chapter 13, and Ms Rahmani, there are few artists working on the sites under discussion in this book. It suggests that there are other avenues of collaboration through which manufactured sites projects can be engaged.

Note

1   The Bioengineering Group Web Site: http://www.bioengineering.com

Chapter 13

Fresh Kills landfill:

the restoration of landfills and root penetration

William Young

Background

This paper introduces the results of field testing of landscape site technologies in landfill restoration. In addition, several myths regarding the performance of tree roots on landfill caps are dispelled through on-site testing, literature research and the observation of precedents in the field. The implications of this research offer scientists, engineers and designers the potential for vigorous habitat creation and restoration for plants, wildlife and people through the reclamation of municipal waste sites.

At the Fresh Kills landfill located on Staten Island, New York, as shown in Figure 13.1, the focus of these field trials, the design team's initial landfill engineering research led to

Figure 13.1 Overview of Fresh Kills site, Staten Island

178

Figure 13.2 Tractor pulling hay mulcher on wildflower experiment

innovative landscape techniques and treatments for final landfill cap and cover treatments, the application of stormwater management, and landfill closure planning. The design team members' mission as part of the New York City Department of Sanitation (NYCDOS) was in the design and engineering of the final cap and cover to the landfill. There was initially a large budget for the final cap and cover because of the team's regulatory mandate as landfill engineers, but some of that money was diverted to support the research. Many of the field experiments were suggested by Andropogon Associates, a landscape architecture firm of Philadelphia, Pennsylvania, using the Fresh Kills landfill as an open air laboratory. These included wildflower experiments, as shown in Figure 13.2, approaches to grass and tree seeding, garbage berms, bioengineering of landfill swales and salt marsh restoration. It needs to be stated that the landfill is not a good place for detail planting, its large scale, and very coarse in texture. In this form of work the design is to be seen as you move at 40 m.p.h., not up close.

Grass and tree seeding

One of the experiments was dormant grass seeding. This method accommodated the landfill capping construction schedule. A seeder attached to a tractor originally developed to restore arid lands – a machine invented by Robert Dixon – was employed on the landfill slopes and performed one-step soil texturing and seeding, as shown in Figure 13.3, that emulated the hoof prints of great plains animals. Instead of having sheet erosion these imprints are created as microsites that stimulated seedling development using low maintenance grasses such as little bluestem, indiangrass and switchgrass, as shown in Figure 13.4. This became the choice seed mix for final cover that was tracked into the soil on steep slopes. On daily cover on the landfill that was not ready for a final cover application the team did most of the seeding. A "ladder" effect was created with a low-ground-pressure bulldozer to help the water infiltrate into the soil rather than run off in sheets. The team would seed first, then track the seed in, going up and down to make the

Figure 13.3 The imprinter machine performs one-step soil texturing and seeding

Figure 13.4 Robert Dixon, the inventor of the imprinter machine, thigh-deep in little bluestem, indiangrass and switchgrass

Figure 13.5 Hydro-mulched seed is tracked into the soil on a steep slope

"ladder" effect, as shown in Figure 13.5, and haymulch over that. The team worked with the Ecology Program from Rutgers University, members of which were interested in working with disturbed lands. In the next experiment, as shown in Figure 13.6, the direct seeding of trees was carried out with half the plots fenced off from herbivores, the other unfenced. A record was kept of what worked and what did not and why, and the data-base built up from there. The fenced plots had better germination and seedling development, but neither plot developed beyond the seedling stage.

Figure 13.6 Direct seeding of trees by Rutgers University

Figure 13.7 Dredge "topsoil" to cap an old garbage berm to screen the landfill

Garbage berms

The team also worked with dredged spoils. A garbage berm funded by the Borough President's office made beneficial use of 30,000 cubic yards of spoil material dredged out of Flushing Bay to expand the Rikers Island Prison. The design team's charge was to use this material to create a planted berm along the West Shore Expressway to screen the landfill from motorists. The team realized that the dredge spoil was soil washed into the New York Harbor, so it had some contamination but was basically a growing medium.

When the material was tested, it contained compounds that were not conducive to planting. To accelerate the remediation process the material was dried, treated with probiotic agents (Huma-Gro) and regularly re-tested. When it tested clean (mostly the high salt content dropped), this topsoil was used to cap and regrade an old garbage berm, as shown in Figure 13.7, that was in a highly visible area from a highway that ran right through the landfill. Normally our approach was to seed first and wait at least a year before the trees were planted. Here, because it was funded from an outside source and had high visibility, we planted right away.

As shown in Figure 13.8, we installed good native plants, communities of mixed-sized local pioneer and mid-succession species that could make it on their own. The establishment period was key for our maintenance to watch out for exotic invasive species, and hopefully it would be self-sustaining after that. There were pleasant surprises along the way, but our strategy was to provide a basic matrix and work with nature. This experiment represented a shift toward making beneficial use of waste products on-site. To be self-critical, the design team did not have the staffing to carry out follow-up monitoring or a maintenance program to develop this highly visible berm into the kind of demonstration showcase that it could have been. The team also carried out bioengineering work on swales that was integrated into landfill design as an alternative to riprap. On a bioengineered swale at the base of Section 2/8, we installed fascines, brush layers and live stakes as shown in Figure 13.9. It is now colonized with sambucus, willow, and dogwood.

182

Figure 13.8 A mixed-sized planting of local pioneer and mid-succession species

Figure 13.9 Experiment on a swale at the base of Section 2/8, with installations of fascines, brush layers and live stakes

Figure 13.10 Salt marsh restoration as an experiment for wetland mitigation

The design team also carried out salt marsh restoration, as shown in Figure 13.10, both as an experiment and for wetland mitigation. The landfill lies on a giant salt marsh, so it is the disturbed and decimated community on the site. The team's work was planting the fringes that were left along the creeks which bisected the landfill.

Demonstration planting

Of all the experiments the most significant was the large-scale planting of trees on the landfill. It was first discovered that there was no long-term budget set up for a 2,400 acre site on part of the landfill. The team considered a self-sustaining natural plant community to be the most appropriate response, and while the team was happy with grassland plantings, they were not the climax plant community in New York. A natural forest community was the outcome of the 55 inches of annual rainfall and good soils found in the region. The "demonstration planting" experiment represented the rare phenomenon of planting large quantities of trees on top of a landfill cap to see what happened.

In 1988 therefore, the New York City Department of Sanitation (NYCDOS) designed, built and monitored a coastal scrubforest on a 6-acre capped area located on the back side of Section 3/4 of the Fresh Kills landfill. The site, as shown in Figure 13.11, on a corner of the landfill was picked because it was on the salt marsh that was remaining on-site and connected into the greenbelt, a band of mature natural lands that is one of the jewels of New York City. At the time, the New York City Dept of Sanitation was not operating under a Consent Order or permit, so the team viewed this as an opportunity to perform work that normally would not be allowed. After seeing the benefits of native grasses growing wild on over forty acres of the capped landfill, the design team sought to demonstrate, through a controlled field experiment, the benefits of using a similar planted diverse, self-sustaining vegetative landfill cover. They selected a closed portion of the landfill overlooking Staten Island's greenbelt, as shown in Figure 13.12.

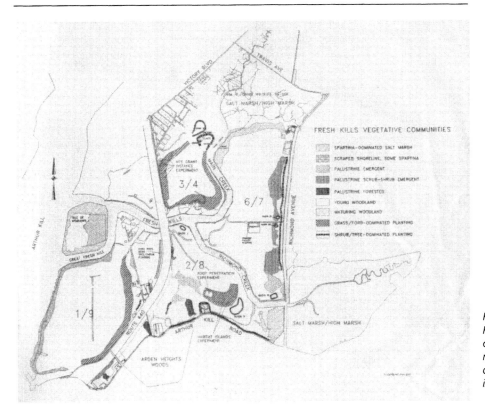

Figure 13.11 Plan of Fresh Kills, showing the creeks, the four mounds, and where demonstration planting is located on Section 3/4

Figure 13.12 View of the 6-acre site

The design proposal developed four basic plant communities, analogs of native plant communities found or once-found on the nearby greenbelt. The clay naturally produces a very low pH soil so these communities included (acid loving) heaths, pine with oak scrub, oak with pine shrubforest and eastern prairie. Turfgrass has up till now been the most recommended vegetative cover on a cap by the US Soil Conservation Service (SCS) and other programs. The design team knew that the eastern prairie with grasses would be more acceptable to the other engineering and solid waste management professionals involved in the landfill than woody plants, so the eastern prairie (primarily little bluestem, switchgrass, indiangrass, broomsedge and forbs) was selected as the major percentage cover over the final landfill area. However, in the remaining parts, smaller 6-acre demonstration plantings of woody vegetation were used and proved to be instrumental for the long-term closure planning and restoration program of the landfill.

The type of community the team was looking to create was a prairie being invaded by various woody plants. The nearby woody vegetation of the greenbelt provided a good analog to the site conditions of the landfill – shallow soils over bedrock and 3/1 slopes. The ericaceous plants included *mirica* and *calmia*, and in these new plantations trees as well as shrubs and grasses were installed on the landfill cap, a practice that went against current thinking in landfill engineering. Tree roots, it was believed, would in time penetrate down through the capping breaking the integrity of the final layer and causing ruptures to the surface. In time, because of the advantageous nature of the tree roots and their network across the landfill, this would lead to the breakdown of the entire capping system.

In 1989, nurseries were not offering the varieties of native plants they offer today. The design team worked with the Parks Department to rescue trees being lost to development. Using a Big John tree spade, unique native specimens a mile from the site were dug up and transplanted, especially oaks, to enrich our diversity. Some of the trees died, but even the remaining root balls provided some new species. From plant rescue we were able to add *Lyonia mariana*, *Quercus illicifolia*, *Q. macrocarpa*, native blueberries, and grasses. The seed dispersion by animals, especially birds, added tremendous diversity in just one year. A couple of years after the planting the gradual encroachment of woody shrubs over the grassland could be seen, as shown in Figure 13.13. The bayberries had allelopathic quali-

Figure 13.13 The gradual encroachment of woody shrubs over the grassland a couple of years after the planting

186

ties that allow them to colonize large areas from one parent plant. After four or five years most of the pines and oaks have put on tremendous growth, with the number of total plant species more than tripling since completion.

Advantages

Many benefits arose from this plantation study; it provided a model that could be applied over the rest of the landfill, wildlife started recolonizing the area and new native (and exotic) plants began volunteering rapidly. The site gained one new free tree for every three that were planted in just one year. Independent monitoring by Rutgers University reported that the tree species count doubled in less than a year. Perching sites drew in birds, whose droppings added new species. The design team had not considered this, and yet it became the core concept to restoring the entire landfill. In addition, due to colonization, we had an entire plant and wildlife community functioning on a capped landfill.

Uncertainty about what the city wanted to do with the closed landfill led to the master plan idea of "habitat islands," as shown in Figure 13.14, to restore the entire landfill on a limited budget. Rutgers University carried out a study to determine how far away these "habitat islands" had to be from the remnant woodland to be effective as targets for the birds to come in and bring new seedlings, and how small could they become. The idea was that in between these planted "islands" the landfill surface would fill in naturally.

In addition to wildlife and increase of species, the public started coming in to enjoy hawk watches and the salt marsh wildlife. The Department (NYCDSOS) showed the New York State Department of Environmental Conservation (NYSDEC) that it could carry out restoration with minimum expenditure. Less than $30,000 was spent on maintenance in the process (shredded Christmas trees acted as mulch, with some replacement planting and one spraying of an anti-dessicant). Finally there was talk of linking this landfill area to

Figure 13.14 View of habitat islands

the 2,500-acre greenbelt, New York's jewel of natural park areas. This also led to a follow-up study, again by Rutgers, of the value of these small demonstration planting areas, or "habitat islands" as they were named, as a restoration tool. The 1,000 acres of the final landfill surface could be colonized at a low cost with a self-sustaining ecosystem.

Disadvantages

The disadvantages were, first, that this plantation was technically unauthorized. In the Part 360 regulations, the planting was difficult to permit through the DEC, if not illegal. This is where the team started hearing the objections to having tree roots over the cap. Second, the design team were not prepared for the number and the proliferation of exotic species, about 5 percent, that rapidly colonized the plantation. In this approach the number of invasive species cannot be predicted. This was a major problem during the transition from viewing the landscape planning and design of landfills as an approach based in restoration rather than horticulture. The design team also became keenly aware of their own poor monitoring skills in the field. The Rutgers University field research addressed that issue and taught the team along the way. As a result of the above, and having to deal with the New York State Department of Environmental Conservation (DEC) due to a signed Consent Order, it led to the team revising how the program for the final landfill cover was going to progress, especially when the DEC regulators were opposed to planting trees on the cap. Very ambitious closure plans hung in the balance, therefore the design team decided to confront the issue head on. NYC Landfill Engineering members met with state officials from the Department of Environmental Conservation to ascertain their objections to having woody vegetation on the cap and what the team could do to prove, once and for all, what the actual effects were. The result of the meeting was that the DEC agreed to a three-year study with three components to ostensibly prove whether roots can negatively affect a clay cap.

Root penetration report

The SCS Engineers and Rutgers University Department of Biological Sciences prepared the report. The components of the study were to be as follows: a site planting to be carried out with a full suite of proposed native plants over a certified cap. Root encroachment of the clay cap was to be monitored. In addition a complete and thorough literature review on vegetation and liners was to be carried out, and excavation of volunteer trees on nearby certified or non-certified landfills in the New York area would take place to see what the tree roots had done to the cap. This methodology was agreed to by the DEC.

In Part One, carried out in 1992, an experimental plantation of seventeen native tree species was installed on section 2/8 of Fresh Kills. Individuals of each of seventeen tree species were planted in a block pattern of 12-foot grids, the same planting was also located off-site on a non-cap area as a control. Rutgers excavated representative samples (1–9 plants per species) in Oct.–Nov. 1992 and November 1993. Fifteen of the seventeen planted species were deemed large enough to be examined in 1992. Rutgers also examined volunteer trees to enrich their data. The DEC accompanied Rutgers in their examination in 1993.

In Part Two, carried out from 1992–4, a literature review was carried out, and over 500 references were consulted. Some were included in their entirety in the Report. The consensus found from examining the nearly 500 references was that 90 percent of tree roots were to be found in the top one meter of soil.

In Part Three, from 1992–3, excavation of existing tree species at other landfills was carried out to determine the progress of the roots *vis-à-vis* penetration or avoidance of the cap layer. At Brookfield landfill, adjacent to Fresh Kills, individual trees were photographed, measured, and excavation was performed every season and documented. The DEC was invited to attend, but often did not come to witness if the roots impacted the cap. The following criteria and methods were used: trees were selected up to eight years in age, the clay cap was examined for roots and root penetration, soil bulk density tests were carried out and soil probes were made with a pocket penetrometer to measure physical resistance.

The myth surrounding trees on landfill caps is that the architecture of the roots mirrors what is above. What we found from all the studies examined is that there is a threshold of a top 0.6 meters in which 90 percent of the wide-ranging advantageous roots reside. In addition, another myth related to the performance of different roots in the soil – the heart, tap and plate roots. What was found was that the roots conform to the environment, rather than to genetic predisposition, such as a tap-rooted tree when encountering a clay cap will divide into fifteen different roots and spread laterally.

By the end of the study, after a year and a half, no breaches to the cap were found. In all cases, off-cap trees grew deeper roots than the on-cap trees; the roots stopped short of the clay cap, but grew further down on the off-cap plot. The overburden on the cap site ranged from 60 cm deep to over 80 cm deep. It was found that root distortion increased with the bulk density of the soil.

A key study in this regard was from the United Kingdom: *The Potential for Woodland Establishment on Landfill Sites* by M.C. Dobson and A.J. Moffat. This was a landmark document because it included all of the professionals and agencies related to this issue; these included foresters, solid waste consultants, the regulators, government officials and agencies such as the Minerals and Land Reclamation Division, Wastes Technical Division, the Government's Department of the Environment with assistance and input from the Forestry Authority Research Division and National Association of Waste Disposal Contractors. The study concluded that bulk density is the key factor to how roots develop. As the density of the soil, the compaction, and the loss of aeration takes place, there is a reciprocal change in root structure and morphology. At 1.2 gm/cm the roots grow fairly unrestricted throughout the soil, yet at 1.4 gm/cm distortion started to become evident. At 1.6 gm/cm the roots started turning back upward and had less fibrous branching. At 1.8 gm/cm the roots became completely distorted and were unable to penetrate the soil beyond a shallow depth. A well-engineered clay cap has a recommended bulk density of 1.8–1.9 gm/cm. It is like concrete, with a pH of 2.5, and no self-respecting root would want to go into that. The typical value of pressure required to puncture or tear a 2.5 mm thick high density polyethylene liner is about 24 MPA. This is approximately ten times greater than the maximum pressure that can be exerted by a tree root (Dobson and Moffat 1993). For example, a volunteer gray birch (as shown in Figure 13.15) dug up from Brookfield, a landfill closed under pre-RCRA standards and not even correctly capped,

Figure 13.15 Photo of a gray birch dug up from Brookfield landfill, a landfill closed under pre-RCRA standards

grew roots that exhibited startling deviation from genetic traits by confining all its roots to the thin mantle of overburden over the non-certified cap, even though the clay was poorly compacted and the soil cover was no more than six inches deep. In short, it grew like a pancake. Similar results were found in excavations on local landfills throughout the New York metropolitan area.

Again, the study by M.C. Dobson and A.J. Moffat that was referenced in the literature review of the Root Penetration Report concluded that the restrictions on tree planting on capped landfills was unwarranted. The study stated: "since 1986, flexibility in landscaping restored sites has been considerably reduced as a result of guidance contained in the Department of the Environment's Waste Management Paper No. 26, Landfilling Wastes. This recommends that trees should not be planted on containment landfill sites." As a result of the Dobson and Moffat study that concluded that there is no empirical evidence that roots breach caps, and that tree roots do not exert enough pressure to penetrate an engineered cap, the directive contained in the DOE's Management Paper No. 26 excluding trees from containment landfills was rescinded. Trees are now more commonly planted on caps in the United Kingdom, and a large number of parks and Festival Gardens with immense tree planting programs are located on former landfill sites. To date, the DEC has not rendered a formal decision on the Root Penetration Report, and the restoration program has not been adopted by the New York City Department of Sanitation.

Landfill forests?

In the landfill restoration field, the landscape architect usually comes cap in hand at the end of the process. The design team first became involved as landfill engineers and then moved into their role as landscape architects. One thing is certain, there is a paucity of real restoration on capped landfills in this country. This includes the application of tree massing over broad areas of the final cover. At so-called restored landfills small clusters of trees, if used at all, are still confined to deeper-soil areas. If the EPA has no problem with trees on the cap, why are there not more forests on landfills?

Bibliography

Dobson, M.C. and A.J. Moffat (1993) *The Potential for Woodland Establishment on Landfill Sites*, London: Department of Environment.

Robinson, George R. and Steven N. Handel (1992) *Narrative Report on the Habitat Island Concept*, Biology Department, Rutgers University, October.

Robinson, George R. and Steven N. Handel (1993) "Forest Restoration on a Closed Landfill: Rapid Addition of New Species by Bird Dispersal," *Conservation Biology* 7(2) June.

RESPONSE

Horticultural research at the interface of remediation, waste management and site design

Perhaps more than any other case study in this compilation, this project best illustrates the relationship between research and innovative practices. While many models have been presented that describe a strong relationship between research and practice, the Fresh Kills project may come closest to the laboratory approach, in which pioneering remediation methods and systems are carefully monitored so that data can be generated to support future uses of these strategies for landfill capping, stormwater management and habitat development. What's particularly interesting in this project is how the innovative thinking and applications grew out of both resource constraints and technological limitations. The garbage berms resulted from the chance availability of local dredge materials as a beneficial waste product. Another resourceful strategy was the use of salvaged local native plants that created the opportunity to "work with nature" as the landfill was reforested. Clearly as engineers and designers explore unchartered territory, their reliance on the skills and resources of researchers, particularly from local academic institutions, becomes all the more significant when building up a database and assessing benefits, both predicted and unanticipated. In the case of Fresh Kills, two unanticipated benefits were the increased bird populations and the relative low costs of the restoration techniques.

In addition to the ongoing studies done during this project by researchers at Rutgers University, prior research was notably used to inform and validate innovative strategies. William Young and his associates established precedence to various studies from England, that resulted in the rescinding of the existing policies and enabling large-scale revegetation to be implemented over the engineered cap. Young notes, as many conference participants have done, how important the interdisciplinary nature of the research and review process was to the Department of Environment that had jurisdiction over the site, because it included a diverse range of professionals and government officials.

This landmark project represents a benchmark of remediation project as laboratory. Many of the techniques, including the use of seeding machines, the "laddering" of the soil to increase infiltration, the planting of trees on the engineered cap, the use of planted "habitat islands" to begin the process of reforestation were methods rarely used in this country for landfill reclamation. As a precedent, Fresh Kills landfill represents an achievement of synthesis between engineered and ecological systems. Several other projects in this book present the great potential of plant material for remediation, but in this case the intent was to recreate a native forest not for soil remediation but instead for control of stormwater run-off. The benefits at this scale are enormous because the site in its fully reforested state represents a considerable expansion of the existing open space/greenbelt system. The many benefits of an enlarged native habitat ironically result from the conversion of a refuge dump into a wildlife sanctuary. As William Young states, the team went through a process of transformation evolving from engineering to landscape architecture as the reforestation was implemented. This notion of transformation, and reliance on interdisciplinary teams, echoes what others have implied, including Eric Carman who suggested that the next evolution in phytoremediation would be to add landscape design professionals to the project teams. Clearly in created wetlands, phytoremediation, brownfield redevelopment and now in landfill reclamation, the application of environmental engineering in combination with landscape design offers greater benefits than each discipline alone would produce, including reduction or elimination of contaminants, containment of contaminants, redevelopment, created habitats, increased open space, recreational opportunities and regenerative systems often providing treatment at a lower cost than artificially engineered solutions. If we are to reap the full benefits of engineering, science and design, it is, as this paper so well illustrates, imperative that those involved in these projects collect the critical assessment data so that others may use this information to gain the endorsement of the governing and permitting agencies.

Chapter 14
Crissy Field:
tidal marsh restoration and form

Kirt Rieder

This essay questions the definition of "restoration," relative to a landscape that has previously been wholly transformed and is physically unrecognizable from historic records. Crissy Field was a tidal marsh for hundreds, perhaps thousands of years, and was filled to become an airfield for a century. Crissy Field is currently being transformed back into an ecologically functioning marsh. This essay also questions the presumption that restoration is achieved only when a "pure," or total, restoration takes place according to one tightly exercised definition, rather than an interplay between a return to nature and the cultural layering of a site that balances and steers the restoration. Cultural layering in this instance refers to the combination of current public and recreational use in an urban context, combined with the selective identification of a period of historical significance by the National Park Service (NPS), to tell the story of a particular interval of time. This process begins to favor specific intervals of time to the exclusion of others, and raises the issue of balancing current use with a proposed reuse of the site driven by historic precedent.

Crissy Field is a sprawling, near flat, 100 acres situated on the northern waterfront of San Francisco's 1,600-acre former military installation, the Presidio. The Presidio, as shown in Figure 14.1, was continuously occupied by military forces, Spanish, Mexican and Ameri-

Figure 14.1 Presidio aerial, with non-conforming Crissy Field runway and structures, c. 1993

Figure 14.2 View along Crissy Shore, and proposed location of the tidal channel.

can, from 1776 to 1994, when the US Sixth Army vacated the post as part of the ongoing nationwide base closure effort initiated in the early 1990s. The Army had long permitted public access throughout the base, especially along the Crissy Field shoreline, leading to the base of the Golden Gate Bridge as shown in Figure 14.2.

The transformation of "post to park" in 1994 has increased public access and the initiative to reclaim Crissy Field as one of the single largest public open spaces in San Francisco.

The Crissy Field project is currently under construction in 1999 and 2000, a manufactured site transforming a degraded military landscape into a public open space landscape. NPS archival research of historical Army records and documentation reveals a lengthy history of sweeping physical changes to the landscape and built structures of the site (Rieder 1994). The recent post to park transformation from military fortification to open space resource hints at further physical changes to come under NPS stewardship.

The technology that has shaped the site in successive waves is particularly well documented and plays an unusually elevated role in the design process for giving physical form to a naturally sustaining tidal marsh (Haller 1994). Current technical site investigations illuminated the constraints of restoring a portion of what was once an extensive tidal marsh, and has in turn strongly influenced the design. The parallel restoration of a culturally significant grass military airfield on a portion of the tidal marsh further constrained the ability to restore the marsh to the pre-military configuration, to an idealized "natural" condition. The grass airfield geographically occupied a portion of the earlier tidal marsh, posing a quandary in the effort to restore both for the NPS.

Crissy Field, as shown in Figure 14.3, is situated on the western portion of what was originally an extensive back dune tidal marsh stretching from near the Golden Gate throughout what is today the Marina District. By the early 1800s, the Army began to transform the vibrant marsh into a flat, logistical staging ground. The shallow water immediately north of what is now known as Crissy Field provided the first protected ship anchorage just inside the Golden Gate, but was separated from the coastal terrace of the main post of the Presidio by 500–600 feet of marsh (Haller 1994).

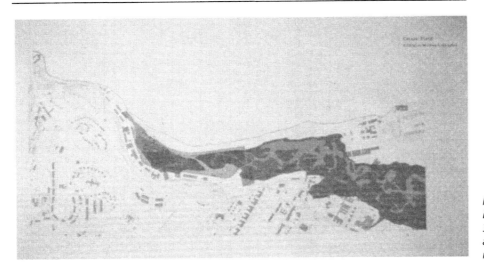

Figure 14.3 Diagram illustrating overlay of 30-acre historic airfield and 180-acre tidal marsh

The Army began a multiple-decade effort to eliminate the marsh by erecting pile-supported boardwalks or wharves traversing the marsh to the anchorage. These piers were succeeded by earthen causeways that effectively compartmentalized the marsh, terminating the tidal exchange. The compartments of brackish water were systematically filled in with bay mud and sand hydraulically pumped in from adjacent shoals by dredges and barges. The site was graded to a near-flat datum ten vertical feet above sea level, erasing all traces of tidal mudflats and marsh. The newly reclaimed land would in part be utilized as a logistics staging yard, shooting range and polo field (Haller 1994).

The grass infield of a wooden plank, kidney-bean-shaped automotive racetrack erected for the 1915 Panama–Pacific World Exposition, situated on this newly reclaimed waterfront, doubled as an impromptu landing strip for barnstorming biplanes. The Army would continue this use into the First World War following the demolition of both the racetrack and exposition, establishing the first military air station on the west coast. Archival research conducted by the NPS Cultural Resource staff, particularly historian Steve Haller, identified 1919–25 as the period of significance; the interval of time during which the most historically significant and documented accomplishments took place, and the interval during which most of the remaining airfield structures were built at Crissy Field (Figure 14.4) (Haller 1994).

Figure 14.4 Oblique aerial view of Crissy Field, c. 1919, showing wear patterns on grass airstrip

Figure 14.5 Crissy Field concrete runway and chain-link fencing prior to construction, 1994

Crissy Field served as a base for numerous aviation record flights, witnessing the setting of distance and time flights, as well as development of technological advances such as aerial photography.

The heavily trafficked grass airfield soon gave way to a longer, asphalt-paved airstrip to accommodate the requirements of more powerful aircraft in 1936. The last fixed-wing aircraft departed Crissy Field in the mid-1970s. The marsh and sand dunes were effectively stabilized, entombed behind rubble and beneath asphalt periodically placed by the Army to armor the beach against erosion (Rieder 1994).

The NPS aimed to change this landscape condition in order to restore, to the maximum extent possible, the naturally functioning systems of the pre-military backdune tidal marsh (GGNRA 1994). The shoreline portion of Crissy Field was irrevocably deeded to the Golden Gate National Recreation Area in 1988, affording continuous public access to a deteriorating military shoreline of rubble and chain link, in the spectacular context of the Golden Gate and San Francisco Bay (Figure 14.5).

The National Park Service and the Golden Gate National Parks Association (GGNPA) initiated several planning studies in the late 1980s, culminating in the General Management Plan of 1994, anticipating the acquisition of the entire Presidio. Coincident to the sudden announcement of an impending Army departure, the GGNPA, the non-profit, fund-raising affiliate to the broader Golden Gate National Recreational Area, retained Hargreaves Associates to provide a physical design for the NPS master planning goals and objectives previously identified for Crissy Field (Figure 14.6).

Prior to construction of the park, and as a condition of the agreement of the transition from post to park, the Army agreed to clean up the contaminated areas of the Presidio, including Crissy Field. With nearly a century of military occupation, the Army had contaminated the Presidio with a variety of toxics. At Crissy Field, petroleum hydrocarbon plumes from aircraft fuels and various motor pools dot the site (Macluff 1998). Additional plumes of pesticides, as well as solvents from cleaning aircraft, military vehicles and tanks, have been documented.

Figure 14.6 Illustrative site plan representing final design of airfield and marsh relationship, as well as final channel location, 1998

Montgomery Watson and the IT Corp concluded remediation of several contaminated sites, through one of three strategies (Macluff 1998):

- Excavate contaminated soils and leave massive holes to be dealt with as part of the park construction.
- Excavate and haul away severely contaminated soils for incineration offsite, replace with native soils from within the Presidio to avoid introducing non-native seed stock and soils to the park.
- Excavate soils for treatment by Low Temperature Thermal Desorption (LTTD) on-site and replace.

Low Temperature Thermal Desorption (LTTD) was used to extract organic contaminants from the soil, by cooking or heating the soil in a large mobile kiln to a temperature of 600–700 degrees fahrenheit. Desorption is the reverse process of absorption. LTTD is roughly an 8- to 9-year-old process, still considered innovative but no longer experimental, and is relatively common. The kiln rides on an eighteen-wheel trailer, and is fed contaminated soil by a front-end loader. The vapor is collected and redirected into a catalytic oxidizer that converts the vapor to H_2O and CO_2, with residual dust gathered in a bag house. The dust is reinterred in the next batch of soil, diluted and reduced to non-detectable levels (Figure 14.7).

Figure 14.7 LTTD facility at Crissy Field, with excavated soils under blue tarps, 1998

The LTTD unit can process six months of excavated soil in less than six weeks, with contaminated soils excavated and stored in large soil bins, awaiting the arrival of the mobile kiln. Around 7,000 tons of contaminated soil were processed by the Presidio LTTD in 1996. At Crissy Field, 380,000 cubic yards of soil were reconfigured; a majority excavated from the marsh to be repositioned on the airfield. Of the total 380,000 cubic yards, less than 10 percent of the soils were processed by the LTTD unit (Macluff 1998). Contaminated soils and groundwater plumes were documented at several locations across Crissy. However, these areas were not proposed for excavation, given their proximity to historic buildings, and therefore have not proven to be a limiting constraint to the design.

With site remediation complete, the question of what the restoration should look like returns. The quantitative and prescriptive technical guidelines proposed by the various engineering disciplines engaged in the restoration leave broad interpretation for generating physical form for a vast sustaining system (Porter 1998). One often prevailing solution is to discount the current urban context and cultural use of the site in favor of a complete return to an idealized vision of nature. A more appropriate alternative, though nonetheless controversial, is to synthesize a technically functioning marsh that accommodates the recreational users and the anticipated visitors for educational and interpretive programs. Furthermore, is it even economically feasible to consider a comprehensive or *complete* restoration?

As one example of many design decisions confronted, the location of the tidal channel was examined from several vantage points: historical, political, technical and design (Figure 14.8). The proposed restoration of this crucial component, the sustaining element of the design, would, over the course of four years, be tested in numerous locations (Dames & Moore 1994). The channel became the pivotal point for the project in that without a well-positioned channel, the reduced acreage of the marsh would be prone to frequent closure and mounting maintenance, and therefore non-sustaining or natural (Williams and Josselyn 1994).

The historical documentation of early photographs and navigation charts clearly illustrated that the "naturally evolved" tidal channel occurred off-site of Crissy, further east

Figure 14.8 Illustrative site plan blow-up, showing early channel location, 1996

Figure 14.9 Channel grand opening connecting tidal marsh and San Francisco Bay, 1999

and south, centered in the Marina District (Rieder 1994). The only remaining remnants of the marsh are the pools adjacent to the Palace of Fine Arts, outside the Presidio. There was no feasibility in proposing restoration or re-establishment of the known/functioning channel through an intensely wealthy and developed neighborhood.

Politically, the channel had to "look" like and function according to the publicly held perceptions as to what a natural channel should look like, ruling out a hydrologically superior concrete culvert. This was central to the effort to provide educational and interpretive opportunities for the NPS, as well as securing donor support. Similarly, the vocal constituency of boardsailors extolled the virtues of this particular stretch of beach as one of the best boardsailing locations in the world. The protected nature of the beach, situated behind an existing groin, offered the best chance for reducing closure of a tidal channel as well as maintaining current boardsailing conditions (Figure 14.9).

From a technical standpoint, hydrological and marine engineers advocated for the same protected location coveted by the boardsailors. The team marine engineer was coincidentally an avid surfer who helped the design team understand both viewpoints. Historical documents, coupled with current statistical analysis of littoral transport, indicated that the channel would function to a greater level of confidence in the protected location, without generating an abundance of decaying plant matter that the boardsailors feared. The proposed 20-acre marsh, a reduced portion of the once 180-acre marsh, would avoid unacceptable risks only with a protected channel (Williams and Josselyn 1994).

Geotechnical investigations reinforced the view that the placement of rubble flanking the pipe groin afforded the best chance for keeping the channel stable in place, as well as functionally operational. The combined technical data of the various disciplines conspired, in the end, to relocate the channel to the optimum location to ensure a high probability of continuous function (Figures 14.10 and 14.11).

As a restoration, the size, location, configuration and projected stability of the channel are markedly different from that of the "naturally" occurring tidal marsh that evolved and periodically shifted, known only through historical photographs and charts. Given the obvious contradictions between what is proposed relative to what was known through documentation or speculation, is it accurate to consider the proposed Crissy Field a "true" restoration from any vantage point?

Are there projects of this scale, in similar urban contexts, that meet the idealized definition of restoration, or is it perhaps more appropriate to conceive of restoration projects such as Crissy Field as a manufactured or re-manufactured site?

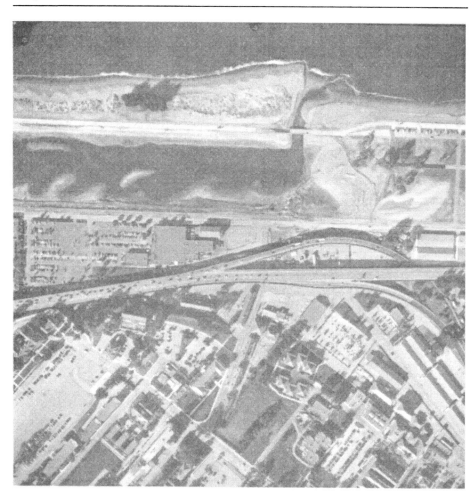

Figure 14.10 Aerial construction photo of tidal marsh, 1999

Figure 14.11 Crissy Field tidal marsh prior to channel opening, with tidal groundwater, 1999

Summary

Crissy Field is a case study in conflict and the resolution of conflict through the design process. Natural restoration and cultural restoration to the "era of significance" became issues for two reasons. The definition of restoration was organizationally defined by the National Park Service (NPS), but not necessarily embraced by the public. Similarly, the geographic positions of both the marsh and airfield during different eras, partially occupying the same location, defied a full restoration for both, given that they are near-polar opposites of each other. In this critique, the marsh restoration serves as an example for commenting on how the definition of a restoration was influenced.

The restoration of Crissy Field was broadly conceived of as a transformation of a recently defunct military waterfront into the centerpiece of a new national park. As the design developed, presentations to both the NPS staff and during public meetings increasingly hinged on the "restoration" aspect of both the natural and cultural landscape, as illustrated in paired "before and after" figures (see Figures 14.12–14.19). The earlier NPS General Management Plan specifically called for restoration at Crissy Field, and the feasibility of restoring the tidal marsh was soon after bolstered by a technical hydrologic study (GGNRA 1994). The full ramifications of executing the restoration of both cultural and natural resource components of the landscape where less well explored.

Figure 14.12 "Before" photo of tidal marsh looking northwest, 1996

Figure 14.13 "After"
rendering of tidal marsh
for same location, with
interpretive overlook,
1996 (rendering
courtesy of Chris
Grubbs)

Figure 14.14 "Before"
photo of tidal marsh
looking northeast, 1996

Figure 14.15 "After" rendering of tidal marsh for same location, with accessible boardwalk, 1996 (rendering courtesy of Chris Grubbs)

Figure 14.16 Rough grading of tidal marsh, with tidal groundwater already a draw for shorebirds, 1999

Figure 14.17 Rough grading of tidal marsh, with boardwalk in place, 1999

Figure 14.18 "Before" photo of restored airfield, 1996

Figure 14.19 "After" rendering of airfield overlooking tidal marsh, reconciling seemingly opposed program uses, 1996 (rendering courtesy of Chris Grubbs)

Bibliography

Dames & Moore (1994) Draft – 90% Document Review, Wetland and Riparian Restoration Feasibility Study, Presidio of San Francisco, June 24.

GGNRA (Golden Gate National Recreation Area) (1994) *Creating a Park for the 21st Century, from Military Post to National Park. Final General Management Plan Amendment*, Presidio of San Francisco. Report prepared for the US Department of the Interior, National Park Service, Denver Service Center. Report prepared by the National Park Service, Denver Service Center, Branch of Publications and Graphic Design.

Haller, S.A. (1994) *The Last Word in Airfields. A Special History Study of Crissy Field*, Presidio of San Francisco, California, San Francisco: National Park Service.

Macluff, Eliana (1998) Montgomery Watson, Presidio soil remediation consultants. Phone conversation (March).

Mead, Aaron (1998) Philip Williams and Associates, Ltd., Crissy Field hydrology and tidal marsh restoration subconsultants. Phone conversations (March).

Porter, Brad (1998) Moffatt Nichol Engineers, Crissy Field marine and civil engineer subconsultants. Phone conversations (March).

Rieder, Kirt (1994) Hargreaves Associates, Crissy Field project lead and landscape architects. Visits to Presidio archives, with Steve Haller, yielded original maps and photographs documenting subsequent transformations (November).

Williams, P.B. and M. Josselyn (1994) *A Preliminary Design Plan for a 20 Acre Tidal Marsh and Shoreline Restoration at Crissy Field*, Report prepared by Philip Williams and Associates, Ltd, San Francisco, for the GGNPA (Golden Gate National Parks Association).

National Park Service (1997) *Cultural Resource Management Guideline*, Release No. 5.

RESPONSE

Natural processes, cultural processes and manufactured sites

The United States Army created Crissy Field by draining and filling tidal (salt) marshes along the San Francisco coast. This process was unusually well documented through photographs and other data. Historical research by the site's new owners, the National Park Service (NPS), established a framework for the redesign of Crissy Field into public open space. The Natural Resources Division of the NPS mandated the restoration of the marshes that had once existed on the site, and the Cultural Resources Division of the NPS required that the historic grass airfield be restored. The fact that these two types of landscapes had not existed on the site at the same time was not to preclude both restorations. This paradoxical mandate set up a challenging problem for the designers: although the problem was framed, what criteria would guide their work?

As Rieder's essay points out, how the term "restoration" was defined (and by whom), and the design implications inherent in a chosen definition of "restoration," was an ongoing part of the design process. The most often-used definition of restoration by ecologists is the one from the National Research Council (1992): "the return of an ecosystem to a close approximation of its condition prior to disturbance."

According to "The Secretary of the Interior Standards for the Treatment of Historic Properties 12," quoted in "Protecting Cultural Landscapes: Planning, Treatment, and Management of Historic Landscapes" by Charles A. Birnbaum (http://www2.cr.nps.gov/tps/briefs/brief36.htm), "Restoration is defined as the act or process of accurately depicting the form, features, and character of a property as it appeared at a particular period of time by means of removal of features from other periods in its history and reconstruction of missing features from the restoration period."

Crissy Field's new twenty acres of marshes are defined by ecologists as a partial restoration as they reflect that a portion of the site's original ecosystem has been restored. Is the airfield a restored cultural landscape? It does rebuild the airfield in the exact same place, and to the exact scale that it was originally, although it was rebuilt as a landform. The marshes were not on the site when the airfield was, which calls into question how the airfield meets all the criteria set forth in the guidelines. These ask that features from other periods of history be removed and missing pieces from the restoration period reconstructed. It seems that the airfield is also a "partial restoration."

Definitions of "nature" are key to the physical form of this project. The essay states that "Politically, the [tidal] channel had to 'look' like and function according to the publicly held perceptions as to what a natural channel should look like, ruling out a hydrologically superior concrete culvert." Who is defining what is "natural"? In the case of Crissy Field's channel, "natural" may have been equated with self-sustaining: the original marsh evolved over time. Historical photos provide evidence that the original tidal channel shifted its position. The new channel is not "natural" in that it is permanently fixed in place, although a natural process – tidal exchange – is facilitated by this channel. Fixing the channel in place was necessary to provide the highest likelihood of a successful tidal exchange, which is critical for the long-term life of the marsh.

The essay highlights the issue of scale as an important aspect of restoration projects. The Crissy Field marshes were part of a much larger hydrological system (a "natural" boundary), one far beyond the legal boundaries of the current site. It was impossible to restore the tidal channel in its original location, but this does not preclude restoring the site if we look back to the ecological definition of restoration. A new channel location has been selected to create the optimum conditions for a functional tidal marsh. The scientists on the team determined that twenty acres of marsh was the minimum amount needed to be restored to assure the health of the marsh. So while the channel form and location were debated, the necessary size of the marsh appears to have been more scientifically quantifiable.

The decision that the channel had to look "natural" was linked to public support of the project, and especially to financial support from donors. What is "natural" meets expectations of what a "restoration" should be. "Natural" may imply esthetically pleasing forms and plantings deserving of political and financial support. A culvert was considered the highest guarantee for success of the all-important tidal channel in a long-term way, while the "natural" channel was more subject to getting filled in. This raises interesting questions about a society's value system: is it impossible for a culvert to be esthetically compelling? Is human involvement considered a part of the site's ecology? Or is human involvement solely a cultural intervention?

The interest in a "natural" tidal channel did not seem to translate into a mandate for "natural" dunes. Within the marsh itself, the new dunes are a sculpted, abstracted nature. They reveal that they are human-made, and therefore not "natural." Again, it depends on one's definitions. The human-made quality of the dunes seems appropriate; the whole site was completely altered in the past, and now it is being remade again into a public park.

"Culture" appears to be defined as human use of the landscape in this essay – which argues for a more expansive view of the role that cultural systems can contribute to park design. At Crissy Field, this might have meant that a channel created by human hands could be expressive of that, and while the design team could have creatively worked with this scenario, another layer of decision-makers determined that the public perception of a park could not be expanded to the same point. Rieder reveals an interest in a more complex look at the notion of "restoration," "nature," and "culture," advocating that present uses of the site are equal players in a restoration, and that this can be a part of the healthy functioning of an ecosystem.

Crissy Field engages definitions of "restoration," "nature," and "culture" to create a multi-layered design that will expand visitors' perceptions of what a park can be. The project remakes both natural systems and cultural artifacts by utilizing an innovative language of built form. The form of the marsh, the "natural system," will reveal that it has been constructed through human intervention. The airfield landform, the "cultural" artifact, will allude to its history as a working airfield, but its function is denied. The airfield has been remade into a sculpture about the airfield. The marsh may also be considered a sculpture about a marsh, but since it is designed to facilitate the natural process of tidal exchange, functioning without human intervention, it is both "natural" and "cultural." Perhaps the remade nature of the Crissy Field site, and the process taken to redesign it yet again, argues for collapsing the dichotomy between natural and cultural landscapes. By encompassing paradox, Crissy Field becomes a rich palimpsest, revealing a powerful approach to the redesign of sites that have been altered, disturbed, remediated, and need to be remade once again.

Chapter 15

The Sydney Olympics 2000:

combining technology and design in the planning of the "Green Games"

Michael Horne

Introduction

This paper concerns the development of two urban landscape technologies that help realize the design vision for the Sydney Olympics 2000 site at Homebush Bay. Specifically, the paper examines the development of structural soil and "eco-paving" for a major planting of *Ficus hilli* defining the eastern edge of Olympic Plaza. The paper also describes how these technologies connect, conceptually and functionally, with the site's overall water management system, via constructed wetlands, creating a manufactured system. In doing so, the paper identifies key issues associated with the development of these technologies, emphasizing the benefits of a collaborative team response to the challenge of remade landscapes.

Background

The Master Concept Design (MCD) was developed by Hargreaves Associates landscape architects and the Government Architect Design Directorate for the Olympic Co-ordination Authority, beginning with a two-month workshop held at Homebush Bay, in late 1996. The HA/GADD team has maintained an ongoing design coordination role, spearheading design and technical issues ahead of the documentation and construction teams.

The MCD binds together an urban core of 300 hectares comprised of the major Olympic venues, exhibition and commercial facilities rising phoenix-like from the former Metropolitan Meatworks site, as shown in Figure 15.1, abandoned in the 1960s to become a *terrain vague* used indiscriminately as a dumping ground for Sydney's waste. The urban core is surrounded by the Millennium Parklands, an emerging mosaic of remnant woodlands, remade waterways and residual armaments bunkers.

Three key design "moves" are proposed within the MCD, giving form to the public domain. The Red Move is Olympic Plaza, the central urban space of the site, over 9 hectares in size, as shown in Figure 15.2. The Green Move has five components: a central park called "The Overflow"; the Fig Grove defining the high point of Olympic Plaza; a series of "green fingers" linking Millennium Parklands to the urban core; an Urban Forest of scattered eucalypts framing both the Stadium and Olympic Plaza; and a line of figs defining the eastern edge of Olympic Plaza. The Blue Move is the use of water, at the Fig Grove and at Haslams Pier, celebrating connection to the wetlands, creek and river system beyond.

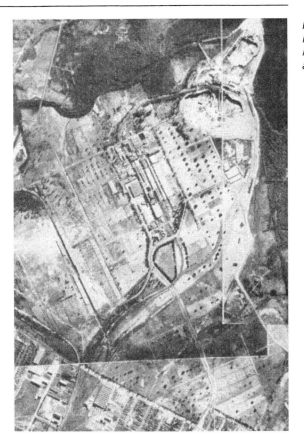

*Figure 15.1 The former
Metropolitan
Meatworks site
abandoned in the 1960s*

*Figure 15.2 Olympic
Plaza, the major
pedestrian area within
the site is lined on one
entire side by a double
avenue of ninety Ficus
hilli*

At the line of figs in Olympic Plaza, these design moves overlap, functionally and conceptually, creating a "manufactured" system – the Plaza figs rest in eco-paving which harvests rainwater, delivering air and moisture to the structural soil below. Excess water within the profile is collected and diverted to the Haslams Pier wetlands, part of the site-wide water management system. Thus, the three design moves are linked: trees (green) to plaza (red) to water (blue).

The Plaza figs

The Olympic Plaza figs form a continuous green edge along the eastern side of the Plaza, creating a densely shaded walk linking the Olympic Park railway station to Haslams Pier, a major recreational feature of the site. Our initial concept design proposed abattoir figs for the Plaza – fully mature remnants of the earlier Metropolitan Meatworks which had been transplanted and stored on-site for incorporation into new works. However, our investigations revealed a number of conflicts relating to their use in the Plaza – the root ball consumed too much pedestrian space and their low branch habit posed problems for pedestrians.

With this in mind, we investigated the planting of new figs, exploring issues such as overall pattern, spacing, canopy management and tree pit size. The final layout proposed ninety *Ficus hilli* in a grove configuration, two lines of trees, 5 meters apart at a 10-meter diagonal spacing, each tree in a 5-meter diameter tree pit. Tree size was also an important consideration, in terms of providing initial shade and also in maintaining a sense of scale with surrounding buildings.

Super-advanced, nursery grown trees were tendered as the preferred size. Fortunately, the successful consortium offered 8- to 12-meter-high trees sourced from a tree farm in far north New South Wales. The trees were then assessed *in situ* by our OCA assessment team. Were they healthy and of consistent form? How had the roots developed and in what soil type? Was branch structure sound and would branch habit conflict with expected crowd movements? What impact would transplant operations have on long-term viability?

The team concluded these trees best satisfied the visual and functional requirements identified in the Master Concept Design. So began the complex tree transplant operation – from the tree farm, by road to Tweed Heads, then by barge 600 kilometers down the New South Wales coast, up Sydney Harbour and into Homebush Bay.

Structural soil

During the fig procurement process, soil volume emerged as a significant issue. What was required below-ground to optimize the long-term potential of the Plaza figs? The conventional tree pit surrounded by highly compacted subgrade/sub-base materials is not an optimum condition for tree growth. While the tree pit, if properly designed, may adequately meet the tree's short-term requirements (air, water, and nutrients for root development), there is no provision for root expansion beyond this immediate zone. Many trees fail to reach their full potential when forced to exist under such conditions. The absence of a useful soil profile beyond the tree pit can also encourage adventitious root development nearer the surface, at the interface between sub-base and pavement, where water and air can more readily be found. This can result in pavement heave and cracking.

The simple solution is to increase tree pit size, hence soil volume, but in Olympic Plaza, as in many other urban situations, this was impractical, mainly due to demands of pedestrian movement and so on. Another solution is to provide a continuous soil trench. Pedestrian demands are met by bridging the soil trench using a structural system of piers and reinforced concrete, on which the selected pavement system is constructed. This approach tends to be technically complex and very expensive. On the positive side, the available soil volume can be significantly increased.

During our team's investigations, we became aware of a paper by Patricia Lindsay, then at the University of California, Davis, that summarized *engineered* or *structural soil*, particularly the work of Grabosky and Bassuk at Cornell University. This concept seemed to resolve many of the issues concerning soil volume, but the available literature indicated "product development" was still in its infancy.

We convinced OCA to form a working group to test and develop the product from "first" principles. The working group began as a core team comprised of landscape architect, geo-technical engineer, soil scientist and arborist concentrating initially on structural soil then, later, eco-paving. As product development moved through concept to laboratory testing, field trial and implementation phases, the group broadened its range of expertise. This highly collaborative, interdisciplinary approach proved instrumental in accelerating product development from research to implementation in the space of four months.

Structural soil is a pavement sub-base, which is also a growing medium for trees. The mix consists of rocks of a single grading. These rocks bear on one another, providing structural stability for the pavement above. As they bear, voids are created – this is where the soil mix resides. The aim is to fill each void by 50–70 percent, minimizing "cushioning" of soil between rocks, whilst providing space for air and future root expansion. Structural soil is also permeable. That is, it is designed to transmit water and air through the profile, turning the prevailing engineering approach to pavement design on its head. However, it must be stressed that structural soil does not replace the soil in the tree pit – it provides a secondary resource beyond the tree pit, under the surrounding pavements. The "filler soil" resides within an inert rock matrix – therefore there is less soil by total volume than provided by *in situ* soil, although this loss is offset by a well-designed filler soil, the presence of air and water throughout the profile, and good management practices.

Structural soil was developed over a number of phases, each involving intensive testing and evaluation from design, engineering and horticultural perspectives.

Phase One involved review of available literature and understanding the fundamental design principles and issues. Several mixes were prepared, testing variations of aggregate size and type, as well as filler soil options. Basalt (40 mm aggregate size) was found to best meet engineering criteria due to resistance to weathering.

Phase Two laboratory trials involved full-scale profiles of each trial mix prepared in glass test boxes, as shown in Figure 15.3. Mixes were subjected to flushing (twenty bucket

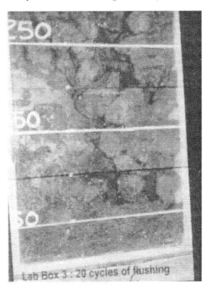

Figure 15.3 Laboratory box showing the condition of structural soil test mix after twenty cycles of flushing with seven liters of water

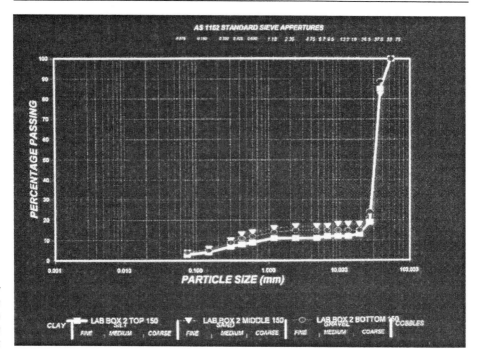

Figure 15.4 Laboratory box 2, graph illustrating particle size to percentage passing results

loads), testing rates of particle migration and suspension of fines through the profile, as shown in Figure 15.4. The trial mix containing a polymer binder was found to offer minimal reduction in particle migration. The preferred filler soil comprised a sandy loam to minimize clay content, dolomite for improved cation exchange, and a small amount of composted green waste for biological activity. The polymer binder was considered unnecessary, given its predicted short lifespan.

Phase Three comprised a field test plot constructed to test buildability issues. The plot was constructed to a draft specification using material batched on-site. Installation was similar to traditional subgrades (i.e. 150 mm layers using six passes of a vibrating roller). Only minor variation/separation of the mix was evident at the surface. The specified tolerance of +/− 40 mm was achieved. A tree pit and trench excavation showed stable, vertical angle of repose to the trench walls.

Rigorous geotechnical analysis, including plate load compaction and nuclear densometer testing, was carried out by independent laboratories. The structural soil profile was found comparable to a Class B pavement (i.e. able to accommodate pedestrian traffic). Despite the positive test results, the lead contractor was not able to guarantee the paving, due to the inherent risk of an untried product. OCA accepted the risk and instructed the contractor to proceed with documentation and implementation.

Phase Four implementation by the head contractor used a work method statement based on the findings of the working group. Denton Corker Marshall, Landscape Architects then documented the extent of the structural soil/eco-pave zone and subsoil drainage system, as shown in Figure 15.5. Installation comprised a continuous 10 meter wide × 700 meter long × 500 mm deep "trench" of structural soil layed over prepared subgrade (complete with subsoil drainage), followed by installation of a 100 mm blinding layer of 2–20 mm crushed basalt over the structural soil. This blinding layer enabled a true surface for the 50 mm deep bedding layer of the 2–5 mm crushed gravel

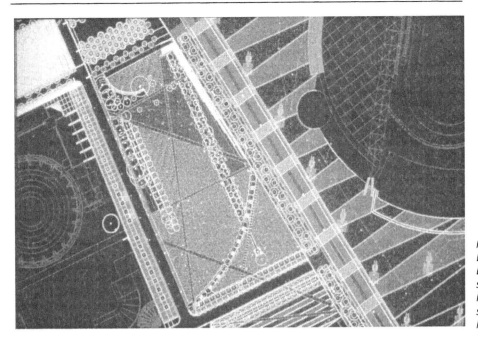

*Figure 15.5
Documentation by
Denton Corker Marshall
showing the Plaza fig
layout within a zone of
structural soil (light-blue
line)*

ready for eco-paving. The working group conducted regular testing of the material during this period.

Subsoil drainage, irrigation system and various services for the Plaza were then installed. Trenching using conventional machinery presented no real problems, although the material was difficult to hand dig. The material maintained a good angle of repose in excavated trenches and tree pits. However, the blinding layer described above was easily scuffed by construction vehicles maneuvering over the material, requiring minor reworking and compaction of disturbed areas. Tree pits were then excavated, the Plaza figs installed and backfilled with the specified tree pit soil mix. Excavated material was stored on-site and reused.

The product development process brought to our attention a number of important considerations for both our team and for future designers. These involved matters such as air and water, drainage, buildability, installation, ongoing maintenance and quality control.

The long-term success of structural soil requires adequate air and water throughout the profile (i.e. under the pavement): in the case of standard paving applications, significant below-ground infrastructure such as a slotted pipe/drip irrigation network, layed mid-profile with surface openings as required. This approach is dependent on effective maintenance and an understanding that the irrigation is not essential until roots grow beyond the tree pit, whereas permeable paving harvests surface water, offering a sustainable, low-energy solution that minimizes costly below-ground infrastructure. "Eco-paving," which is discussed below, was developed in response to these considerations. However, use of this paving type does involve some loss of pedestrian comfort, which may be unacceptable in high-intensity pedestrian zones such as a city sidewalk or where a dedicated accessible route is planned. In these areas, the former approach may be necessary.

Subgrades at Homebush Bay are highly compacted with low permeability – a situation common in urban tree plantings. Irrespective of the approach to aeration and irrigation described above, it is necessary to ensure adequate drainage at the subgrade interface to prevent waterlogging and harmful anaerobic activity. The subsoil drainage system under the Plaza figs intercepts excess water and diverts this to the constructed wetlands of the Northern Water Feature.

Tree grate footings must be designed to minimum depth in order to maximize contact with the surrounding structural soil. Similar care should also be taken with header course footings and edge restraints. Structural soil also involves a significant additional cost over and above standard pavement sub-base. At Homebush Bay, this was partly due to the pioneering aspect – the work method was unfamiliar and the product unknown. However, costs are likely to fall as acceptance of the product increases and market forces influence supply costs. Nonetheless, the cost of structural soil will remain more than conventional sub-base.

OCA accepted liability (after an exhaustive process of product development) for what was an untried product in Australia. Already, structural soil is being specified for other projects. Designers should ensure that the client is aware they are using an innovative product with only a short track record; however, the issue of liability is likely to recede as structural soil is seen to perform in a variety of situations.

If properly constructed, structural soil requires minimal ongoing upkeep except in situations where subsurface irrigation and aeration is in place. In this instance, more intensive maintenance will be required.

The next step in development of structural soil is ongoing monitoring and evaluation, in order to assess properly the long-term performance of the material in regard to soil chemistry, root development/penetration, particle migration, drainage, and related issues. Procedures such as controlled trials and destructive testing are yet to be implemented at Homebush Bay.

Development of understood standards is fundamental to the long-term success of structural soil as a generic product. Development of Australian and American Standards for structural soil could greatly assist this process. Quality control is important in three key areas: specification, batching, and installation. It is essential that soil experts routinely audit batching and installation specifications. The contractor should also prepare a specific Work Method Statement for installation of structural soil, to reinforce understanding of installation issues. Batching and shipment to site is critical. For example, excessive or poorly distributed filler soil in the batch may create areas of "spongy" bearing material which could lead to slumping of both the structural soil subgrade and the pavement. This sort of failure will ultimately undermine acceptance, and hence use of the product.

"Eco-paving"

The Red Move in the Master Concept Design is Olympic Plaza, a paved pedestrian area of 9.5 hectares uniting the major stadia and the new Showgrounds. The adjacent Overflow park contains significant existing trees and creates a new "Green Heart" – in a symbiotic relationship with the Plaza. The Plaza is intended as an open forum able to accommodate large crowds. Shade is provided by the buildings themselves, and along the eastern edge by the Plaza figs. Nineteen towers, designed by Tonkin Zulaikha architects, provide signage, seating, shade and lighting at regular intervals within Olympic Plaza. Solar panels mounted on cantilevered awnings will collect energy and feed back into the power grid. In essence, the Plaza creates a big rug, a vast welcome mat, its pattern generated from the juxtaposition of the new Boulevard and old abattoir geometries that organize the site.

The Master Concept Design team recommended an interlocking concrete "tri-hex" paving system for Olympic Plaza based on esthetic, performance, and life cycle considerations. This system was rigorously prototyped in a series of field trials testing issues of paving color, detailing, finish and the like. During these trials, our team learned of Eco-loc, a modified interlocking pavement manufactured by Rocla, designed to trap and drain

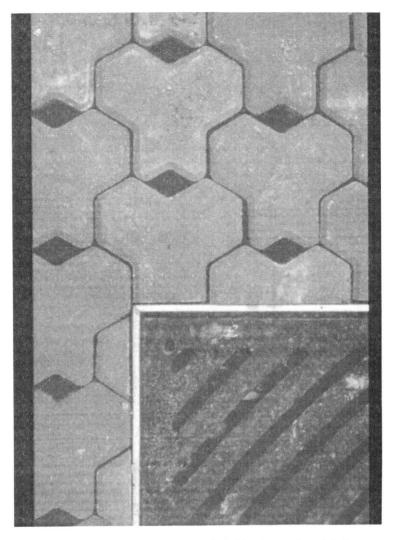

Figure 15.6 Completed eco-tri-hex paving

water. A sample of this product was subsequently included in the paving trials but was found to be unsuitable for pedestrian use.

However, the principle was sound, so we worked with Rocla to develop a new "eco-paver" shape as a modular component of our preferred tri-hex paving system. Rocla responded by producing a small sample panel of the first eco-tri-hex prototype, hand cut from standard tri-hex pavers then installed in the paving field trial to test pedestrian comfort and pattern issues. This eco-tri-hex shape created a perforated field of paving, harvesting rainwater via a surface of small drainage holes, each filled with 2–5 mm gravel. Preliminary calculations indicated a water infiltration rate equal to a ten-minute storm with a recurrence interval of six months (Sydney conditions). Further eco-tri-hex were cut and laid over the structural soil field trial to test buildability of the 2–20 mm blinding layer and 2–5 mm bedding layer. The bedding layers and paving, as installed by the contractor, proved to be a relatively simple procedure. Engineering analysis indicated the eco-tri-hex/structural soil paving system was structurally adequate for light traffic load of 2 × 105 ESA over twenty years. The paving system was approved by OCA and is now installed, as shown in Figures 15.6 and 15.7.

Figure 15.7 Plaza figs surrounded by eco-paving

The product development process brought to our attention a number of important considerations in the future use of "eco-paving."

At Homebush Bay, we have only used "eco-paving" in areas considered as secondary pedestrian routes. The drainage holes create a textured surface, with a regular pattern of small openings, that was determined as not fully accessible to those with disabilities. However, future work might explore the use of eco-paving as a tactile delineator; for instance, the tree zone in pedestrian sidewalks.

Eco-paving (and structural soil) needs to be well understood by site operations and maintenance personnel if it is to function effectively for the long term. This requires an effective transfer of knowledge from design team to maintenance team concerning the full range of issues that might impact on both pavement quality and tree performance, including:

- A replenishment program for the gravel-filled drainage holes is required. Anecdotal evidence from Germany suggests these holes block up with surface matter (dust, rubbish, etc.) and should be replenished every five years. This will require vacuum removal, probably with specialist machinery.
- In addition to long-term blocking, the collected surface matter also requires periodic cleaning, particularly removal of cigarette butts and the like. Periodic top-up of gravel in the drainage holes will also be required due to settlement and loss from cleaning work.
- Pavement cleaning work must also be specified so as to prevent flushing of deleterious cleaning agents through the eco-paving into the structural soil root zone. Conversely, eco-paving may also be used to flush in desirable additives such as fertilizer for the Plaza figs.

216

Conclusion

The Olympic Co-ordination Authority supported development of structural soil and eco-paving as part of its commitment to fostering innovative and sustainable development in the transformation of a degraded industrial site into the centerpiece for the Sydney 2000 Olympic Games. This commitment goes hand in glove with OCA's strong focus on the development of Homebush Bay as a long-term legacy for Sydney as much as for the Olympic event.

The urban landscape innovations described here were made possible by highly skilled specialists working as a team. The landscape architect's role was to bring together the various technical inputs and spearhead a solution that would best meet the design intent.

In conclusion, rainwater is harvested by eco-paving into the structural soil below, providing air and moisture to the root zone of the Plaza figs.

Excess water is collected and diverted to the constructed wetlands at Haslams Pier that are an integral component of the site's Water Management System, as shown in Figure 15.8.

Figure 15.8 Olympic Boulevard under construction, June 1998. The detention pond in the foreground was constructed as the Northern Water Feature, designed by Hargreaves Associates

References

Grabosky, Jason and Nina Bassuk (1995) *Progress on the Development of an Urban Tree Soil Specification*, Ithaca, N.Y.: Cornell University, Urban Horticulture Institute.

Lindsey, Patricia (1994) *The Design of Structural Soil Mixes for Trees in Urban Areas*, Growing Points, Davis: University of California.

RESPONSE

Environmental site design as public infrastructure?

The Sydney Olympics, the "Green Games," is reclaiming a post-industrial wasteland with a comprehensive system of environmentally responsible, sustainable technologies. The use of structural soil and "eco-paving" are innovations that unite the landscape systems operating on the site. An immense plaza that facilitates the huge numbers of visitors has been designed to address the basic biological needs of its trees, while integrating excess water into cleansing wetlands. This integration of urban tree planting research with the site's water management system is a prototype for biological systems and urban infrastructure working together to enhance city life.

The typical condition, however, for trees in the city is that they die at an alarming rate; in the United States the average life of a street tree is seven years. It is only recently that the reasons for their short life span have become known. These trees die because their basic biology has not been well understood; consequently, tree planting technologies have rarely started from the tree's fundamental requirements. In contrast, the Sydney project is utilizing current information to start from the trees' needs for appropriate soil, water, and air.

Research into tree root systems is quite new. The notion that a tree's roots are a mirror image of their canopy has been dispelled: 90 percent of a tree's roots are in the top 24 inches of soil, and a tree's roots spread far beyond their drip line. Recent research developed by landscape architect James Urban and others has concluded that soil volume is the key factor in long-term survival for urban trees:

> The actual amount of soil that the tree needs should be calculated by measuring the area within the projected mature drip line of the tree. A rough rule of thumb is to provide one cubic foot of usable soil for every two square feet of mature canopy volume . . . Studies on the East Coast suggest that providing approximately 400 cubic feet of soil per tree will usually provide sufficient growing room if the other soil in the area has some capacity for root development.
>
> (Urban 1996: 96)

This is a huge departure from the conventional design for trees in pavement which typically shows a 4 foot × 4 foot tree pit, resulting in a tree pit volume of approximately 48 cubic feet. This tree pit detail can work fine for a tree planted in lawn as this tree has adjacent soil to utilize for root expansion. Trees in the city typically have no adjacent source of healthy soil.

The fig trees in Olympic Plaza are in pits approximately 15 feet in diameter – much larger than a conventional design would specify – which promises healthier trees. Although the tree pits are not interconnected, the structural soil is in a continuous trench. This strategy employs recent research that has found that trees growing in interconnected soil volumes grow larger than they would if planted individually. Interconnected soils create the conditions for a more even distribution of water and roots (Urban 1996: 96).

Developing, testing, and implementing the new technologies of structural soils and "eco-paving" at Sydney was the result of a very effective collaboration between landscape architects and scientists. However developing these new technologies is also expensive. The fact that these products/systems were virtually unknown in Australia highlights the importance of the "Green Games" to bring new technologies into the mainstream. The Australian government underwrote the cost of research and development, and assumed the liability for these innovative technologies. It is highly unlikely that any private developer or individual city would have been willing to take on these kinds of expenses and risks. The Olympics provide an incentive for experimentation. The world observes the host country's efforts to create both a setting and an image, in this case, an environmentally responsible one.

The Sydney Olympics utilized the latest data to construct a "manufactured system" for trees to thrive surrounded by pavement. Trees that are used as urban street trees are typically forest climax species, and when these trees are planted in the city they require infrastructure to succeed. It may be argued that the trees themselves are infrastructure, as important to urban life as the utility systems and roads, especially if we consider that plants sustain life on earth by producing oxygen and removing carbon dioxide. Trees provide shade, cooling city streets, and the myriad esthetic qualities of trees – their texture, form, color, and scent contribute profoundly to the quality of urban life.

The line of figs at Olympic Plaza represents the human impulse to control nature. Although technical innovations will support the figs, there is ample reason for their presence. The trees provide much-needed shade along the eastern edge of the Plaza, linking the railway station with recreational centers. The figs serve as an element of human scale, working as a foil to the architecture and the Plaza. The matched specimens will be stunning, but they will probably reach the end of their life span together. Then what will happen? Many would argue to simply replant with new trees of the same species, which is of course viable. *Allées* and groves of the same species are ancient forms of planting design, and it would be a great loss if they disappeared from our cities due to a fear of planting monocultures.

A new paradigm for street tree planting is emerging in Europe which utilizes the idea of plants as systems. In the Netherlands, street trees are increasingly being thought of in terms of an urban forest. Fast-growing pioneer species are planted at the same time as slow-growing climax species. The pioneer species are pruned upward to allow the climax species room to develop. The goal is to have a variety of trees and ages of trees on the site, providing continual canopy in a more sustainable way, while providing an opportunity for city residents to become more aware of the natural processes of growth and succession (Hough Date 1995).

It is valid to see plants as sculptures and to see plants as systems. At the Sydney Olympics, both points of view are visible: a monoculture of mature figs in pavement as well as cleansing wetlands where communities of plants cleanse stormwater of toxins. The "Green Games" provides a new view of urban design, where a city may be built by starting from an understanding of natural processes, then integrating human use into these processes. This synthesis of technology, science, and design is of the utmost importance in the redesign of post-industrial, or "manufactured sites," where remediation of soil and water must be the starting point for design.

References

Hough, Michael (Date 1995) *Cities and Natural Process*, London and New York: Routledge.
Urban, James (1996) "Room to Grow," *Landscape Architecture* (March), pp. 74–97.

Chapter 16
Sydney Olympics 2000:
Northern Water Feature

Kevin Conger

Introduction

This essay will address some of the challenges facing landscape architects and designers who are working with sites and projects that involve remediation or engineering intensive design processes. The case study describes the design process of the Homebush Bay Olympic Site Master Concept Design, focusing specifically on the Northern Water Feature and wetland elements of the project designed by Hargreaves Associates, landscape architects.

The refinement of the design of a project beyond the conceptual phase can present many technological challenges that require the knowledge and expertise of specialized consultants. The success of the project is dependent upon a collaborative process that brings together strong design intent with firm functional, technical, and environmental underpinnings.

Project introduction

The project case study is the site for the 2000 Olympics in Sydney, Australia. The site area shown in Figure 16.1 is 760 hectares (1,900 acres) at Homebush Bay, approximately eight miles west of downtown Sydney, which is at the geographic center of the Sydney metropolitan region.

Homebush Bay was originally mostly mudflats and wetland mangrove forest, but over the last hundred years has been reclaimed by extensive landfilling for industrial uses. The most recent use of the land has been for waste deposit, clay mining from a large open brick pit, and a slaughter yard. The current site is rather flat with some gradual undulation, and is bounded on the north by Haslams Creek, and on the east by Parramatta River and Mangrove wetlands.

Sydney won the bid for the 2000 Olympics in 1993, and since that time the Olympic Co-ordination Authority conducted several intensive efforts to establish goals for the development of the site and create a masterplan. After several different masterplan schemes, a final masterplan was developed in 1994.

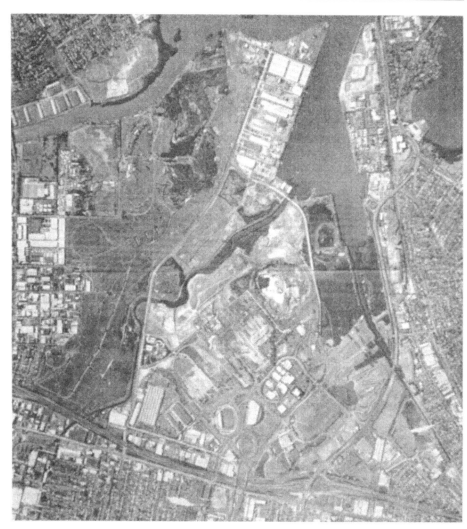

Figure 16.1 Aerial photograph of Homebush Bay site during early construction

Master Concept Design workshop

As the designs of the individual building precincts within the masterplan evolved and in many cases expanded, the areas outside of the building envelopes, referred to as the Public Domain, became fragmented and it became apparent that there was a need to reconsider the importance of a coherent public domain. In the fall of 1996 Hargreaves Associates was selected to work with the New South Wales Government Architect Design Directorate (GADD) to create a new Master Concept Design for the Public Domain. An intensive six-week workshop was held at the project site, resulting in a new concept design that now defines the Public Domain as a distinct precinct within the masterplan as shown in the centre of Figure 16.2.

The new design can be summarized in a series of strategic "moves," as shown in Figure 16.3, Red, Green and Blue. The red move is a major central public space (Olympic Plaza) which unites the buildings and creates a large space appropriate to the scale of the events

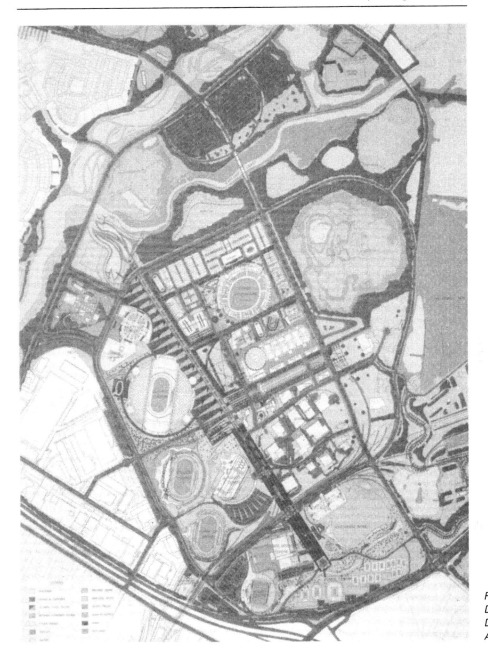

Figure 16.2 Public Domain Master Concept Design by Hargreaves Associates/GADD

and architecture. The green move is the creation of a landscape respite within the urban core, and tree-lined green fingers introduced as the connection of the Plaza to the surrounding parklands. The blue move is the use of water to activate a gathering space at the high point of the plaza and create a connection to the wetland landscape at the low end of the Plaza, the Northern Water Feature as shown in Figure 16.4.

The design of the Public Domain engages the urban core with the larger landscape systems. Connections are made: culture/environment, nature/technology, experience/

223

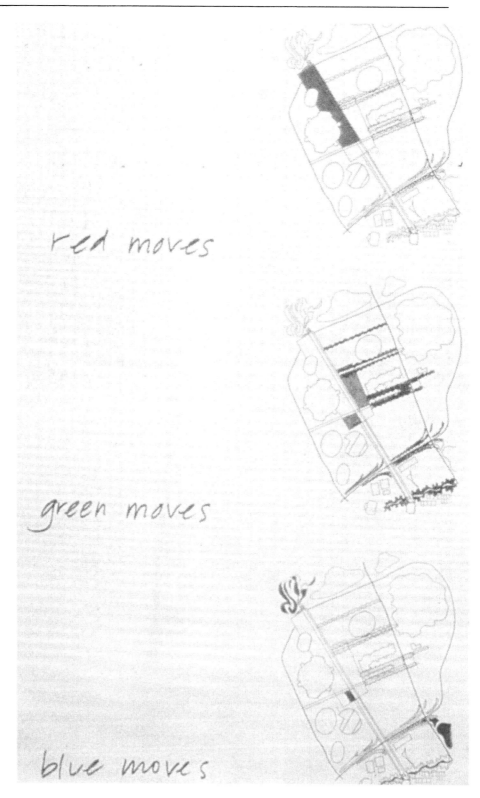

red moves

green moves

blue moves

Figure 16.3 "Red, Green, and Blue Moves" which give identity and definition to the Public Domain

Figure 16.4 Rendering of Northern Water Feature design

natural phenomena, past/present. The Northern Water Feature and wetland component of the design exemplify these connections in an interactive and interpretive water feature that engages the urban stormwater infrastructure and wetland to collect, purify, and demonstrate the water cycle process.

The implementation of the new Master Concept Design involved the reconsideration of the site into distinct and manageable projects. Many local architects and landscape architects became involved in the construction documentation of the different project areas, with Hargreaves Associates developing and documenting the design of the two major water feature areas and continuing to review and coordinate with GADD of the whole public domain.

Site description

The Northern Water Feature site, as shown in Figure 16.5, is situated at the northern edge of the public domain, on the edge of Haslams Creek.

The site is at the lower end of the western stormwater catchment, as shown in Figure 16.6, and contains an existing water quality control pond that retains run-off before releasing the water into the creek. The site is bounded by a capped waste material landfill on the east and west, and roads on the perimeter of the landfills.

Some of the challenges presented by the site were the improvement of the water quality control pond design, enhancement of adjacent wetlands, preservation of endangered frog habitats, and containment of toxic landfills. In addition, the site had to perform an important role as a major public space and create a crucial connection of the public domain to the surrounding parklands.

Figure 16.5 Site of the Northern Water Feature wetland and fountain

Northern Water Feature as part of the Master Concept Design

As the open-ended extension of the Olympic Plaza, the Northern Water Feature is oriented outward to the wetlands and parklands to engage the urban core of the site with the larger landscape systems. The Northern Water Feature is one of the most important components of the Master Concept Design as a gateway from the Public Domain to Millennium Park and Olympic Village, and the point where the Public Domain and nature converge. The interaction of the urban core and the wetlands at this point becomes the physical expression of the environmental awareness that was the foundation of the principles of the

226

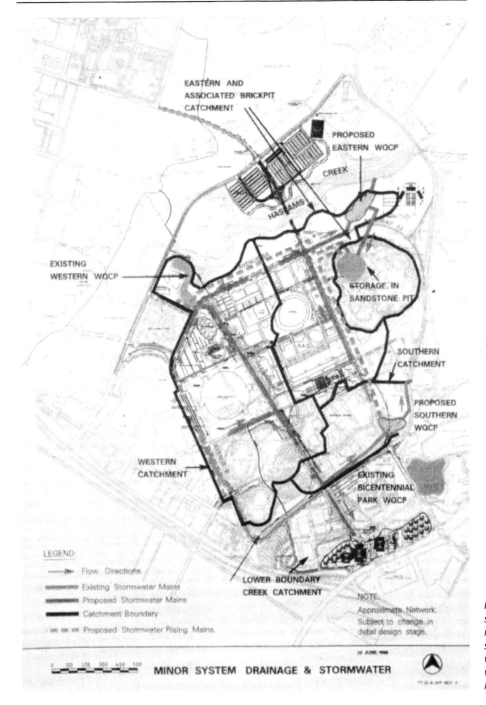

LEGEND:

→ Flow Directions
▬▬▬ Existing Stormwater Mains
▬▬▬ Proposed Stormwater Mains
▬▬▬ Catchment Boundary
▬ ▬ ▬ Proposed Stormwater Rising Mains.

EASTERN AND
ASSOCIATED BRICKPIT
CATCHMENT

PROPOSED
EASTERN WQCP

CREEK

HASLAMS

EXISTING
WESTERN WQCP

STORAGE IN
SANDSTONE PIT

SOUTHERN
CATCHMENT

PROPOSED
SOUTHERN
WQCP

WESTERN
CATCHMENT

EXISTING
BICENTENNIAL
PARK WQCP

LOWER BOUNDARY
CREEK CATCHMENT

NOTE:
Approximate Network.
Subject to change in
detail design stage.

MINOR SYSTEM DRAINAGE & STORMWATER

*Figure 16.6 Site
stormwater
management diagram
showing the 250-acre
western catchment area
which flows into the
Northern Water Feature*

227

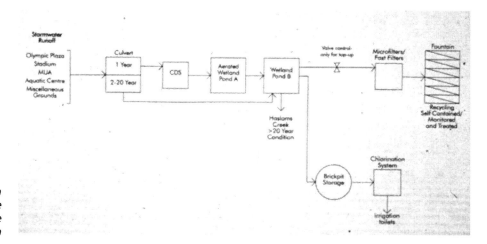

Figure 16.7 Northern Water Feature schematic water cycle diagram

planning and design of the Sydney 2000 Olympic site. The design intent of the fountain was to create a landmark for the lower edge of the tilted plane of the Olympic Plaza. The fountain is composed of three rows of fifteen jets, each jet is shot from 2-meter-tall stainless steel nozzles to create an 8-meter-high arch of water. The result is three grand arching veils of water that step down and outward to the wetland and landscape beyond. The fountain appears to throw water from the Plaza terminus directly into the wetland ponds, connecting the two separate elements into one landscape system. The forms clearly reveal the hand of the designer as part of a designed and constructed site, and are intended to express the flow of water through the wetland ponds. The stone and lawn terraces that step down from the Plaza to the water edge, and the pier that extends outward from the Plaza to overlook the wetlands, allow for interaction and interpretation of the fountain and wetlands.

As part of the site water cycle system shown in Figure 16.7, the Northern Water Feature wetland will receive stormwater from the western catchment of the site, which is approximately 100 hectares (250 acres) in extent. In addition to the stormwater run-off from this area, backwash from the Aquatic Center swimming pool filters will enter the wetland. The site water cycle strategy was to retain the stormwater in ponds before pumping it to a larger storage pond in the brickpit, after which it is treated and used throughout the site for irrigation and toilets. The concept for the Northern Water Feature wetland was to treat the stormwater through the use of macrophyte plant zones in wetland ponds in order to improve the quality so that it can be circulated through the fountain as part of the water cycle. Not only is it important in our minds for the wetlands to contribute to the cleansing of the stormwater run-off, but the demonstration of the water cycle through the wetland and fountain was an important part of forging a connection between the technical and natural forces of the site.

Technical design development

The design of the Northern Water Feature and wetland quickly evolved and changed as the project progressed into the Advanced Concept Design and Design Development phases. The most significant revisions were the result of refinements to the water cycle strategy and how the wetland would function.

Initial design concept options included schemes that incorporated Lemna algae ponds to treat nutrient-rich water sources, presumably from an adjacent sewage treatment facility as shown in Figure 16.8. After more study it was determined that these water sources

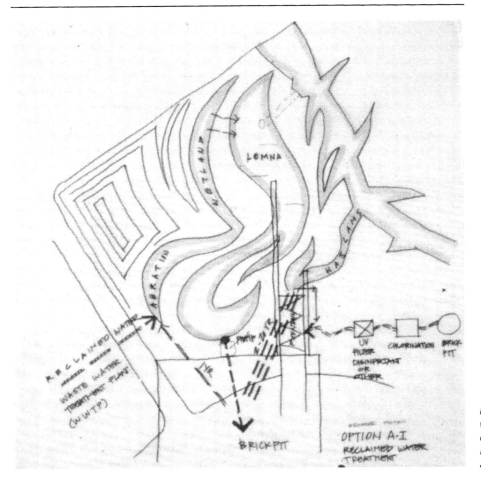

Figure 16.8 Preliminary wetland and fountain diagram showing reclaimed water source and Lemna ponds

could potentially contain bacteria that would require, under Australian regulations, for the ponds to be totally separated from any human contact. Because of the significance of the wetland/water feature as an interpretive element within the Public Domain, we decided it would not be desirable to construct a fence or physical barrier around the entire wetland. Once this water source was eliminated, the bacteria that feed the Lemna were no longer provided and the Lemna was deleted as well. A related concern was that the Lemna was not indigenous to the creek and that, should it enter the creek system, it could potentially be invasive. This adjustment to the water treatment cycle resulted in a significant change in the configuration of the ponds, in that it eliminated the need for a separate aeration pond and allowed us to increase the wetland area for stormwater treatment. Another fundamental component of the wetland water feature that was studied in the early technical design phases, and significantly affected the conceptual basis and form of the design, was the role of the fountain in the wetland water cycle. Initially it was thought that the fountain water would be derived exclusively from the wetland ponds. After further analysis of the stormwater catchment area it was determined that the wetland ponds could potentially receive contaminants that would not be purified by the wetland and fountain filtration systems. Because of the level of public exposure and interaction with the water there was a need to ensure clean and healthy water sources. As a result, it was decided that the water cycle should be modified so that only supplemental top-off water for the fountain would be obtained from the wetland, while the initial filling and

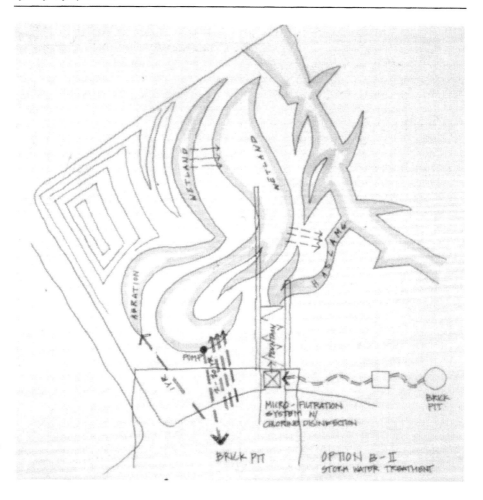

Figure 16.9 Final wetland and fountain diagram

refilling of the fountain would come from a potable source. This resulted in the redesign of the fountain pools to be totally isolated from the wetland ponds, as shown in Figure 16.9.

The final design was based on the following water cycle model, as shown in Figure 16.10. The water from a one year or less storm will pass through a pollutant trap before entering a redesigned inlet pond. The water filters through a wetland macrophyte zone where it is cleansed, and then to an outlet pond before either being "harvested" for site use and pumped to a storage pond for use throughout the site, or pumped directly to the fountain as top-off water. Water is recirculated from the outlet pond back to the inlet pond to maintain constant water flow within the wetlands in order to control algae and mosquitoes. Storms of greater intensity than one year will bypass the pollutant trap and short circuit the wetland ponds to Haslams Creek.

Wetland system design parameters
Once this conceptual diagram is established a series of technical requirements and conditions set up the wetland design parameters. Most importantly are the catchment area, hydrology, and wetland dimensions. The peak flows of the catchment area are taken into consideration. In addition, the flows from the Aquatic Center pool are estimated at a constant monthly discharge of 44ML (11.5 million gallons) (1ML = 264,200 gal.). The critical

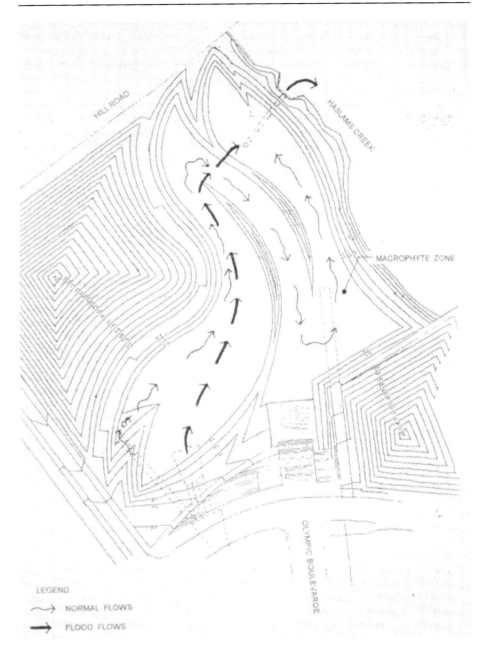

LEGEND

NORMAL FLOWS

FLOOD FLOWS

Figure 16.10 Flow diagram

dimensions for the wetland are determined by the volume of water that is related to the catchment area, the surface area covered by macrophytes for this volume to be treated, and the amount of water level fluctuation that the macrophytes can tolerate. The minimum required volume is 5.25 mil. gal. to the spillway level, and the minimum "first flush" volume required (that is, the amount of water that can be held in the wetland at the beginning of a storm) is around 2.1 mil. gal. In addition, the shape of the wetland should accommodate a range of water depths so that there will be fringing macrophyte growth where the water depth is between 2–3 m, and the proportion of the total wetland

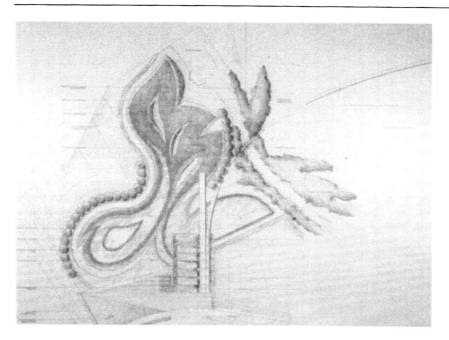

*Figure 16.11
Preliminary site design*

covered by macrophytes is at least 30 percent. The maximum water level variation that can be tolerated by the macrophytes is 0.5 m.

These parameters establish a performance target for the wetland system of 70–88 percent retention in a year of average rainfall. The water quality standard we are trying to achieve is the ANZAC "recreation quality secondary contact," which means you can touch it but it's not clean enough to swim in or ingest.

Wetland design

The application of these design parameters, along with the concept of an interpretive wetland, drove the design of the configuration of the wetland ponds.

The plan, as shown Figure 16.11, illustrates the earlier design showing a larger aeration pond and a somewhat more complex series of forms that posed some circulation problems. The final plan, as shown in Figure 16.12, illustrates a more refined hydrological

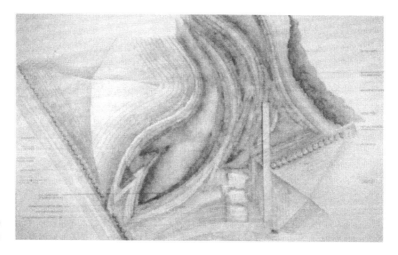

*Figure 16.12 Final site
design*

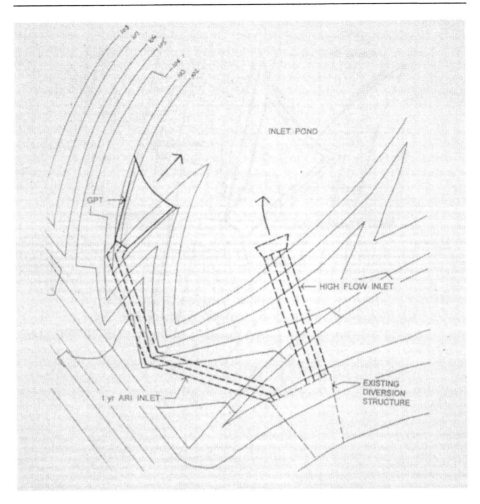

Figure 16.13 Wetland inlet design diagram showing storm event scenarios

model to achieve higher performance. The inlet and outlet designs were modified and, in addition, the forms of the stone terraces changed to be more compatible with the wetland ponds. In order to support this design a number of technical features were added.

Inlet design

The inlet is designed to allow flows up to a one-year storm to pass through a pollutant trap before entering the retention ponds, as shown in Figure 16.13. Flows in excess of one year up to a twenty-year storm will be diverted from the pollutant trap to flow directly into the wetland ponds.

Pollutant trap

An existing gross pollutant trap (GPT), which is basically a low fence in the water that screens large debris and trash, is to be removed and replaced with a new GPT in a new location that is more compatible with the wetland design. A more effective and more expensive type of pollutant trap, a continuous deflective separation unit (CDS), is currently being investigated as a change to the design. The CDS unit, which basically works like the "spin cycle" of a washing machine, forcing trash and large pollutants out of the water, would be less visible and provide cleaner water.

The outlet arrangement of the wetland allows for water to be pumped off to the brick-pit storage pond. The volume of water that can be harvested is 2.1 million gallons, which is related to the first flush volume and water level fluctuation tolerances. This amount of water will take about one week to harvest, based on the pumping capacities. In addition, water will be taken off and filtered for top-off water as needed for the fountain. This could be as much as 3,000 gallons per day in peak season in hot, windy conditions. Water will be recirculated through the wetland in low flow periods by pumping from the outlet pond back up to the inlet pond. This will allow for continual water movement and supply to the macrophytes, as well as allow for multiple water treatment in the wetland.

Bathymtry (circulation)

It is critical for the water to flow constantly and continually throughout the entire wetland pond system in order to reduce algae and limit mosquito breeding. In addition, it is also critical that no water is allowed to "short circuit" through the system within a shorter detention time than is desired for water treatment. The water that enters the wetland during a one-year storm or less will be detained within the macrophyte treatment ponds for about one week.

Flow models, as shown in Figure 16.14, have been generated to inform the configuration of the ponds to design a system that evenly spreads flows throughout the ponds, maintaining velocities that will not damage the macrophytes.

Figure 16.15 is a topographic model indicating water depths. Water treatment occurs in three basic areas of the wetland: a deep inlet pond where the gross pollutants are removed and coarse sediments will drop out, a long sinuous macrophyte zone which is almost completely covered with macrophytes, and a deep outlet pond from which water is pumped out.

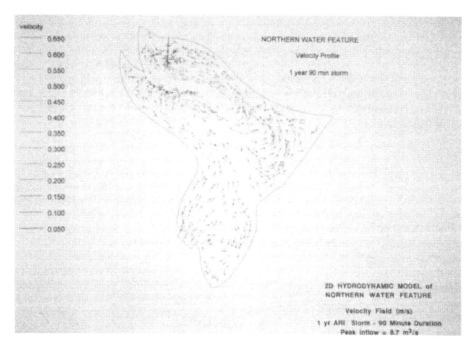

Figure 16.14 Bathymtry models indicating predicted flow patterns

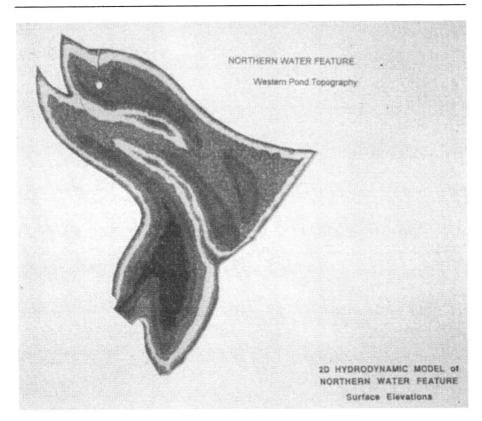

NORTHERN WATER FEATURE

Western Pond Topography

2D HYDRODYNAMIC MODEL of
NORTHERN WATER FEATURE

Surface Elevations

Figure 16.15
Topographic models
indicating water depths

Roughness models, as shown in Figure 16.16, indicate the areas of water bodies that contain macrophytes and those that are open water. The shape and depth of the ponds are designed to provide range of water depths in order to support fringing macrophyte growth at a water depth of two meters. This design achieves approximately 70 percent surface coverage by macrophytes for water treatment. Water level control is necessary within the macrophyte zone of the wetland in order to allow for establishment of the macrophytes without drowning, as well as for maintenance. In this design the water control is achieved with a series of stop board weirs placed between the macrophyte zone and the outlet pond.

The wetland edges are developed to support macrophyte growth and allow for even water flow circulation. The edges are mostly planted, and only engineered where geotextiles, boulders, or gabions are required for hydraulic controls. The gradually sloped and stepped edges allow for maximum plant growth and provide habitat for wildlife, including the endangered green and golden bell frog, and also create a safe zone along the water's edge appropriate for the adjacent public uses. The wetland will appear as primarily grasses and reeds, with some upland scrub on the banks. The higher landforms and slopes will be planted with native grasses.

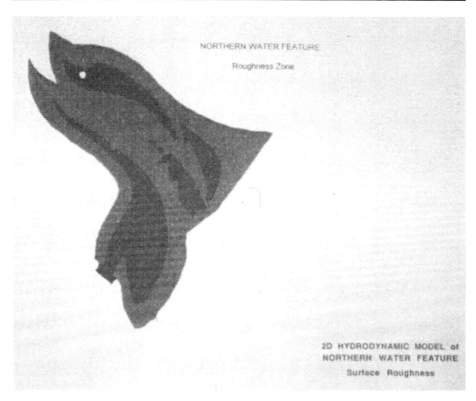

NORTHERN WATER FEATURE

Roughness Zone

2D HYDRODYNAMIC MODEL of
NORTHERN WATER FEATURE
Surface Roughness

Figure 16.16
Roughness model
indicating vegetated
and open water areas

Construction and establishment

As the project award was pending and construction were about to start, there are a few aspects of the design were being tested. The location and identification of all of the endangered frogs on the site, and the relocation of these frogs during construction, was being coordinated by consulting ecologists. This issue was particularly important given the environmental agenda of these "green games." Fortunately, the frogs appeared to be pretty hardy, and seemed to be multiplying faster than the ecologists can count them.

The exact location of the contaminated landfill, shown in the plan in Figure 16.17, was to be determined in the immediate future as mass excavation begins for the bulk earthwork. The containment of any leaching will be achieved with a sub-drainage system. The establishment of the wetland macrophytes will take about three years, and during that period the ponds will be filled incrementally so the plants can acclimatize without drowning. Once the macrophytes are established the water quality will be tested and, when acceptable, be introduced into the fountain system. The images shown in Figures 16.18 and 16.19 illustrate the completed wetlands and fountain.

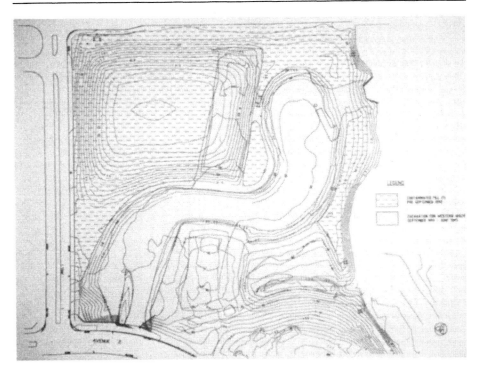

Figure 16.17 Earthwork analysis of existing contaminated landfill

Figure 16.18 View of wetlands with fountain beyond

Figure 16.19 View of fountain with wetland beyond

Summary and conclusion

I would like to return to the issue of collaboration in the design process. There are often many ways to solve a problem. The challenge that designers face is how to take projects beyond problem-solving. With this project our team first developed an idea about what the place should be, and then relied on the technical solutions to inform the design and resolve the engineering issues. Because the design is based on the concept of connections and processes, the forms are able to change and evolve to reflect the technical requirements, making the design more legible and more interesting as these layers are added. A universal appreciation of how this project fits into the bigger picture, what it means, and how it functions culturally as well as technically, directed every decision during the design process. Not only did the technical solutions inform the physical design, but also the design concepts became criteria for evaluating the technical systems.

Although the client for this project had an ambitious agenda, and the project program was exciting and high profile, the challenges associated with the site and requirements of the program are not unique. These conditions exist in many projects, on many sites around the world. Designers and engineers have the opportunity to influence client groups: developers, public agencies and industrial site owners, to consider the potential of these difficult sites as places to be remade, rather than problems to remediate.

RESPONSE

From mechanical to the biological

At the Sydney Olympics, environmental responsibility is stressed; water, soil, and vegetation are seen as systems. Creating a new public domain that enhances the ecological health of this post-industrial site is a primary goal of this project. Not only must the site function for the games themselves, a brief and intense spectacle of a few weeks, it must work in an ongoing way within the city fabric as housing, recreation, and infrastructure. In addition to these challenges, the site contains contaminated water and soil, inspiring the overriding concept of the "Green Games." Cleansing and recycling water through the site is a priority of the project, and the Northern Water Feature is a key component of this strategy.

The design of the Northern Water Feature is the result of an impressive collaboration. The designers' leadership set forth a strong conceptual base; they then worked with the art and science of the project in an integrated way, creating a design language based on the innovative technologies employed. The design of these "Green Games" underscores this shift from mechanical solutions to biological solutions for remediating or enhancing the health of the landscape. The landscape is seen as an ongoing, dynamic set of systems, rather than a place possessing problems that are solved "forever."

The Northern Water Feature is designed to cleanse storm water run-off, and provide a prominent public gathering place. It is important to note that there were several ways that a cleansing wetland could be defined, which in turn would change the physical form of this feature. What water should be cleansed? Stormwater only? And from what parts of the site? What level does the water need to be purified and why? The ultimate form of the feature was established by the design parameters: the definition of what water was to be cleansed and to what level. These parameters were then evaluated for their requirements: water volumes and flows, the requirements of the cleansing plants, and the overall goals of the wetland, which all contributed to the final form of the Northern Water Feature.

It is critical to understand that the designers didn't just solve a water problem; they refined their goals for water treatment and public interaction after evaluating several options. A key question was the relationship between water quality and public safety. The fountain would be enticing to visitors, necessitating that the water be cleansed to a level that would not be a danger to public health. By requesting, interpreting, and generating scientific data, the designers tested and refined their design alternatives.

Typically, water systems are hidden by mechanical infrastructure; the traditional way of dealing with stormwater has been to drain it away as quickly as possible into catch basins. Most citizens have no idea how the water cycle works, or how it is possible that they turn a faucet and water appears. A goal of the Northern Water Feature is to make water more visible, raising questions in the visitors' minds about the role of water in this project, and presumably in other contexts as well.

The Northern Water Feature is designed to provide an educational experience of an ecological project. The phrase "interpretive wetland" is used to describe the design concept, but it is not clear from the essay what an interpretive wetland is or how this concept will be communicated to the public. Was interpretive signage deemed too didactic? Why or why not? Will the fact that the "fountain appears to throw water from the Plaza terminus directly into the wetland ponds" provide enough clues to the visitor that human use and the water cycle are linked both biologically and mechanically in this setting? Inviting visitors right up to the (clearly human designed) Northern Water Feature will provide the opportunity to experience wetlands as part of an urban, public environment, where they can then be seen not just as a piece of "nature" that one sees in the "country," but as a part of city life, and a new kind of "green infrastructure."

In ecological design, problems can be simply revealed to the public or they can be addressed. At the Sydney Olympics, site problems are being addressed and in addition there is interest in this remediation being revealed to the visitor. The Northern Water Feature is designed to be seen as a landscape of remediation, while highlighting the beauty of the natural processes of water movement and plant growth. A powerful addition to the project would be an experimental component, for example, testing one type of plant over another for its cleansing properties. This is the next step that an ecological design project could take.

A precedent for this increased level of collaboration between art and science is the project "Revival Field." Created in 1990, this seminal project was a collaboration between artist Mel Chin and scientist Rufus Chaney of the United States Department of Agriculture. Chaney's pioneering research indicated that some plants had the ability to cleanse soil by processing contaminants through their vascular system. Laboratory research was brought to the field when Chin and Chaney planted a test plot on contaminated land near St Paul, Minnesota. Chin utilized the quartered circle for the form of the garden, an ancient form recalling medieval herb gardens. "Revival Field" provides a model for collaboration between designers and scientists; scientists would benefit by working with more designed landscapes, especially in urban areas, and designers could expand their own knowledge base of remediation technologies as well as contributing to ongoing research.

As seen at Sydney, innovative remediation technologies have implications for a new language of built form in the landscape. The example of the Northern Water Feature shows us that cleansing wetlands – constructed systems that utilize plants to remove toxins from water – can also be urban public space. This is a new and potent type of hybrid landscape, a new type of manufactured site.

Part IV
Postscript

"Dialogues with the Contributors" expands on the research and project work highlighted in the essays in Parts II and III.

Three panel discussions took place on April 4, 1998 during the Manufactured Sites Conference at the Harvard Graduate School of Design. The audience comprised regulators and design and engineering practitioners from the public and private sectors as well as academics and students from design schools across the country. The discussions were led by a moderator and the panel's questions and comments were recorded, transcribed and appear here in an edited form.

The discussions focused on further and more detailed questions arising from the conference presentations, including clarification of new and innovative technical procedures and environmental construction materials, and discussion of interdisciplinary research and design collaboration. The three moderators were Deborah Marton, City of New York Parks and Recreation Department, a practicing landscape architect and lawyer; George Hargreaves, Chairman of the Department of Landscape Architecture, Harvard Graduate School of Design; and Daniel Winterbottom, Assistant Professor of Landscape Architecture at the University of Washington, Seattle. Participating contributors are identified at the beginning of each panel.

Chapter 17
Dialogues with the contributors

Niall Kirkwood

Panel discussion 1

The participants are identified as follows:

Deborah Marton, Landscape Architect, City of New York Natural Resources Group (Moderator).

Wendi Goldsmith, Bioengineer, The Bioengineering Group, Inc.

Lucinda Jackson, Staff Environmental Scientist, Chevron Corporation.

Kirt Rieder, Landscape Architect, Hargreaves Associates.

Steve Rock, Environmental Engineer, US.EPA.

William Young, Landscape Architect, The Dawson Corporation.

Figure 17.1 Panel 1 participants, from left to right, Wendi Goldsmith, Steve Rock, and Lucinda Jackson

MARTON:

What is the efficiency of phytoremediation plants in their dormant period?

GOLDSMITH:

Recognizing the fact that they have a long metabolically active season during the year, the only time when the roots and beneficial bacteria associated there are not active tends to be in the cold period of the year. Chemical reactions in the soil are slowed to a negligible rate; even water movement through soil is slowed. Plants definitely don't work year round but the bacteria that live with plants work longer than plants.

QUESTION FROM AUDIENCE:

How do you dispose of the vegetation after phytoremediation?

ROCK:

Depends what's in it. Different plants in different soils will take the same contaminants up at different levels. Sometimes they will accumulate them in the roots, sometimes they will move them into the leaves, sometimes they won't deal with them at all.

JACKSON:

The beauty of phytoremediation is even if you have a contaminant in the plant tissue, the plant tissue can be reduced to a really small amount of waste.

GOLDSMITH:

In phyto-situations some of our native wetland plants used are giving us 1 percent to 3 percent bio-mass of lead. We've identified a smelter that will pay us for those plants, to actually recycle and recapture the lead. So instead of creating waste product we're reusing it.

JACKSON:

We've had a similar experience with selenium in that you can harvest the selenium out from the plant tissue. Farmers are also interested in purchasing the hay containing selenium for feed for their cattle, because a lot of cattle have a selenium deficiency.

MARTON:

Who does the testing if you need to know what's going on and whether to harvest your plants?

ROCK:

There are companies that now specialize in phyto-extraction. The only two that I know are doing it are Phytotech in Monmouth Junction, New Jersey (now part of Edenspace Systems Corporation of Virginia) and Phyto-works outside of Philadelphia, Pennsylvania. There are also a lot of labs, for example at Rutgers University.

QUESTION FROM AUDIENCE:

In terms of research are government and university research teams simply just planting on secluded sites to see what happens?

JACKSON:

The problem with phytoremediation is that it is very site-specific. Somebody does a test with one plant on one site, somebody does a test with a different plant at another site. So we have a little information on a lot of things. We are trying to pull together all these databases, centralize it so that we can get good information from it. It involves industry, academia and different types of companies who want to collaborate and to share data.

QUESTION FROM AUDIENCE:

Where did phytoremediation come from historically?

ROCK:

If you search for phytoremediation, the earliest article with that word is around 1994 to 1995. Older words such as bio-accumulation and geo-botany were used in the 1930s. The first person to suggest using plants for remediation was Rufus Chaney at the US Department of Agriculture. He noticed that plants were taking up toxic levels of chemicals. He thought: "we cannot use them for eating, perhaps though we can use that extraction capacity for something else."

MARTON:

When should clean-up occur, what are the deadlines, and how do you dovetail the timing of phytoremediation with regulatory requirements?

ROCK:

By the time you get the money together to do a conventional clean-up, a phytoremediation installation will be complete, and will save you money in the process. Part of the answer to why we are here is to discover how to integrate the remediation with landscape design. How to locate the appropriate plants, how to apply them to uptake the contaminants efficiently; so you use the site for a dual purpose – you're cleaning it up while you're using it.

JACKSON:

It takes time to convince the regulators, and sometimes you cannot and we have to do a conventional clean-up. We also can transform the site from a clean-up site to a research site and then they allow us to try an innovative technology.

GOLDSMITH:

When the government itself has financial responsibility for the cleaning activity it is remarkable how much more amenable they are talking about alternative, less-costly, better-integrated solutions with different time-frames.

MARTON:

Landscape architects are trained to look at the landscape as a pattern of patches and the goal in some instances is to increase the flow of biotic elements among patches, increasing connectivity. Is the integration of geomorphology and design a way to address the connectivity of a fragmented manufactured site?

YOUNG:

I believe that we have already over-inhabited the earth and there are far too many people to just think about restoration. In whatever I do, I look at whatever healing processes you can do to complete the little patches that are left. We don't try and recreate nature, we just put in something that will heal whatever the disturbance is.

GOLDSMITH:

I am a geomorphologist by training. Obviously some of the major issues that drive engineering design in phytoremediation projects involve things like the structure of surficial sediments, groundwater patterns and movement of groundwater. Landscape design from a stability standpoint, water movement and ultimately vegetation fit in there rather well. A vegetated slope can be stable at a much steeper angle than an unvegetated slope, for instance. I think if we can understand geomorphology and how it can serve us as a design tool, we can for instance build vegetated slopes instead of structural slopes. These can be more attractive, often less costly and typically far more beneficial biologically for all the reasons that you all have been reiterating.

MARTON:

One of the values that is brought to a site by landscape architects is that the future users should have some understanding of the site history and its former uses. That value may be somewhat at odds with the efficacy of remediation in the sense that it might not be in the best interest in terms of restoration or cleansing the site to have a sign in the landscape of the site's prior use.

ROCK:
The unfortunately prevailing engineering attitude is that we'll get in there and make it safe, and whatever else happens after that is somebody else's responsibility. Engineers have been long accused of focusing, which is the great strength and the great failing of the whole discipline. You put enough brainpower behind a problem and you are probably going to solve it.

RIEDER:
I took the time to talk to some of our engineering consultants for our Crissy Field project and try and get a sense from them of the roles of the consultant over the course of the four years of the process of the project. What did they see their role in restoration of the tidal marsh, did they feel that they were a part of it. It always came back to the word "restoration." I tried to get them to tell me what it means to restore. To a person they said, "we're there to solve a specific technical problem that you have put to us. Take the rubble off the beach, make the channel function."

GOLDSMITH:
I was asked to lead a panel discussion last week at the first conference of its type of the American Society of Civil Engineers on the subject of how should engineers and ecologists work together. Toward the end of the discussion it was clear that engineers and ecologists are not the only consultants that need to be at the table. The other body that needed to be represented is design and planning and that they probably need to lead and coordinate the process. This came as a surprise to the people there. The engineers had not really seen themselves as following directions.

Panel discussion 2

The participants are identified as follows:
George Hargreaves, Landscape Architect (Moderator).
Eric Carman, Environmental Engineer, ARCADIS Geharty & Miller.
Michael Horne, Landscape Architect, GADD, Australia.
Peter Latz, Landscape Architect, Germany.
Sue McNeil, Civil Engineer, University of Illinois, Chicago.
Dante Tedaldi, Environmental Engineer, Bechtel National Inc.

HARGREAVES:
I have a question for Dante [Tedaldi]. Looking at your two overlays, one with the precise mathematical model and one with the smoothing out. Tell me which works when you are presenting this to the public?

TEDALDI:
The Potentially Responsible Party (PRP) refuted the results of the investigation, so there was a lot of skepticism on the part of the public that the problem even existed. In that sense, we had a little advantage in terms of smoothing out the results. What we needed to do was to sell the residents on the fact that there was a problem. Newer residents that had moved in after the plant had closed were interested in the spots where their houses were located. People come up to giant maps with street names and look for where exactly their houses were. That's tricky when they say, my house is here on the red spot and your other map shows that that spot is clean. In the end we had to explain we were developing a feasibility study level estimate of the magnitude of the project. We costed out that 525

Figure 17.2 Panel 2 participants, from left to right, Mike Horne, Peter Latz, and Dante Tedaldi

homes would require clean-up. While at the same time the test approach, I believe, indicated 1,600 homes. The maps helped us to define a cost, is it $50 million, $100 million or $5 million? Using that smoothing method allowed us to cost that.

QUESTION FROM AUDIENCE:
I would like to ask Michael Horne to talk about the differences between the sand and the clay, the polymer, the phosphorus (in the manufactured soil) and how all that may come into play in the overall long-term success of the [Sydney Olympics 2000] project?

HORNE:
The structure of soil is basically a matrix of an aggregate of rocks, butting up against each other for their structural stability. They create voids that are filled up with 50–70 percent soil and the rest of it is air that provides the space for roots to grow into. We were worried that the clay soil in the mix would block the thing up too much, so we went for a sandy loam. A short-term measure was to mix it with composted grain waste. It improves the moisture retaining capacity of the soil for the short term. The nutrients start moving into the soil as the tree roots begin to fill in. The hallmark idea in terms of the research was the polymers. There's no research about whether they hold together in the long term. We came around to the view that it was very expensive in terms of what it was made of; they are not ecologically friendly materials. The conclusion was that the materials could bind enough to be installed and it worked.

QUESTION FROM AUDIENCE:
Will the gravel in the perforated pavements eventually trap plants to grow in them?

HORNE:
They probably will, but it's something that will be managed because it will be a very highly used pedestrian space. The eco-paving works with the structural soil. The structural soil is essentially a gap-graded construction so it's hardly permeable. What we wanted to do was bring water down into the structural soil to feed the trees and provide air and moisture.

The excess water is collected and ends up in the Northern Water Feature, so it does leak back into the overall system.

QUESTION FROM AUDIENCE:

Continuing on the same subject, will we be able to see this process, can we connect this process of water movement from the Great Terrace to the wetland? What I'm missing is how do the technological issues interact with the media to define the spaces that take us through a series of experiences.

HARGREAVES:

In hindsight I think we should have understood that there was a way to express the harvesting of water and where it was going. I was very afraid of the system, the structural soil, the eco-paving, I still am. Now that I see it working, next time I will be more comfortable being expressive with it, making it part of the spatial and experiential quality of the design process.

QUESTION FROM AUDIENCE:

I was enormously intrigued by Mr Latz's project at Duisburg and I was wondering what inspired him to take this very different approach? How did he decide about all the various things that would go on to the site – was it your client, or was it you?

LATZ:

It was very complicated – there were three agencies involved: the town, the state, and the International Building Exhibition as the leader of the project, but really it was a combination of science, art and tradition.

QUESTION FROM AUDIENCE:

Do you think there will be other projects that will be similar to it?

LATZ:

I am working on a new project like Duisburg but surrounded by mines. It presents an interesting approach to what could be the beginnings of an international park just within the borders of three countries – France, Belgium, and Luxembourg – but I think it is really far too complicated!

HARGREAVES:

The reason why Peter's [Latz] project has leftover industrial ruins is because that is the approach that they have been doing in Europe for thousands of years, and it would cost more to remove the ruins than to leave them. In Europe there is more of a tendency to treat what society does as layer, upon layer, upon layer. Whereas here we erase things in order to make them natural. What Peter did was to take it beyond that and introduce a new landscape, one that is above and beyond recent times. I think what Peter has taken on is a celebration of the site, it's a wholly different concept.

MEMBER OF AUDIENCE:

Peter's work is also in response to the materials that he is choosing. I think the difference might be with your work [George Hargreaves] and some of the work that we see done by some of the landscape architects in the United States is that the way that they look into the new material. I think the way you find inspiration for the new work in a new place is the material that you are looking at and using, is this true?

LATZ:

I think it was the materials. The possibility to make a metamorphosis. It was not just to make it feel historical (concerning the industrial heritage associated with the site). It is our heritage and we have to protect it – but that is not possible. In the first stage it is necessary to find out the new hue of these things: how can I use these objects?; then we have the second stage: what is it possible to make in this place?

Panel discussion 3

The participants are identified as follows:
Daniel Winterbottom, Landscape Architect (Moderator).
Kevin Conger, Landscape Architect, Hargreaves Associates.
Richard Owen, Civil Engineer, Arup Environmental, London.
Jean Rogers, Environmental Engineer.
Lorna Walker, Environmental Engineer, Arup Environmental, London.

WINTERBOTTOM:
What is the relationship between research and the implementation being carried out in remediation technologies?

OWEN:
In the United Kingdom clients are looking for proven technology solutions. Our clients are very reluctant to go with innovative solutions because they are not sure if it is going to work and they are not sure of the time-frame. I think there are interesting technological ideas and concepts, but the research and the demonstration of the research has to take place, you have to have one or two projects that do succeed.

ROGERS:
There has to be a balance that is struck between moving ahead on innovative technologies and relying on the tried and true. In the United States when sites are subject to the process that's dictated by the EPA there is a procedure you would go through to evaluate these technologies against another, and it's very explicit what the benefits and risks are associated with each. There is a process of weighing and understanding what works best for each site and there is definitely a reluctance to use an innovative technology. A lot of times there is a real risk associated with these sites, a health risk, to the ecological system, something has to be done quickly, and you don't have the time to play around, and see whether it's going to work. Innovation is a luxury in many cases, if the time-frame is such that they cannot experiment, and someone is in immediate risk on-site

Figure 17.3 Panel 3 participants, from left to right, Kevin Conger, Richard Owen, and Jean Rogers

CONGER:

Related to that is the client's perception and the public's perception about what site remediation is all about. Few of the projects we saw alluded to the fact that the final site design had been remediated. There was no way to interpret that the site had been remediated. What comes closest to that are the sustainable companies that use it as marketing and capitalize on the fact that they are making this investment in positive environmental practices. If companies involved in site remediation were to take that attitude, where they are going to be explicitly demonstrating that remediation in their final design, they will get a more positive image.

As designers we first look to the engineers to choose the functional solutions so that we can then try and reflect those solutions in the final physical design. I think a really good example is Steve Rock's small park where he had proposed one scheme and the National Park Service had another idea about what it should look like. The potential for form – making results from an interpretation of the site that reflects the engineering and remediation processes. What these forms are, designers never know, because they do not know enough about remediation. We try to collaborate, bring people in that will show us perhaps the solution and that will make it evident in the final design.

WINTERBOTTOM:

How do you see remediation as establishing a new framework for land development, conservation, infrastructure, recreation, and is there going to be some form of transformation about how people view these as opportunities? In the Salford Docks Project fish were introduced; could you introduce habitat?

WALKER:

We did have to do a lot just to bring food and oxygen back into the water to get the fish back. There was another quite apparently simple piece of work that was very difficult to do, which was to locate where vegetation could occur in terms of water depth. We did not have all the research, a lot of it was initiation to see what you can do.

ROGERS:

I think it depends on who the client is, and whether they want to make it expressly related to the remediation process or not. If it is a municipal or government client they are more willing to make a "monument" out of it. If it is an industrial client, particularly if it is an operating facility, they want it to be as discrete as possible. They really don't want any memory of the fact that they may have degraded the landscape.

CONGER:

They want to cover it up, and make it look picturesque and pastoral. That's the crime right there, they make so much money on these sites and then they just cover it up and walk away. I don't think it's just enough to do that. One of the biggest constraints is time. There is I think the potential for design form to derive from the time-span of a remediation process or a landscape design that responds from one phase of clean-up to another.

OWEN:

In the multidisciplinary team, it's not only the timing of the project but the timing of when you bring on consultants into the team, for example landscape architects. So you get the balance between something that the engineer is sure is going to work alongside the correct design expression.

CONGER:

The flip-side of the point is sometimes the landscape architect does not understand the processes and pays them lip service and makes something that looks like an engineer designed it, and maybe is not. The cooperation has to work the other way, because landscape architects do not have all the knowledge necessary.

ROGERS:

That goes back to the education process, which is so important. I don't think it is really being addressed in landscape programs, where people are understanding what kind of situations they may face on these contaminated sites, how they are typically remediated, what the suite of technologies is that apply, and how long it will take. They don't get involved early on in the remediation process when these are being discussed. Multidisciplinary teams are generally formed later, but they are not as effective as they could be. The designers are not willing to get in and try to understand the science and work with the possibilities.

WALKER:

The group that we work in, we have biologists, chemists, engineers, planners, environmental scientists and landscape architects and sociologists – we all sit in the same building and work together. I think that pushes the design process further because you have people to test things off – surely that is the way it has to be, particularly with the complexities of manufactured sites.

Figure credits

Cover image, Latz + Partner, image courtesy of Christa Panick; Chapter 1 figures, Niall Kirkwood; Chapter 3 figures, Chevron Corporation; Chapter 4 figures, ARCADIS Geharty & Miller; Chapter 5 figures, Steve Rock; Figure 6.1, Sue McNeil; Figures 6.2, 6.3, 6.4, Sue McNeil, courtesy of Engineers Society of Western Pennsylvania; Chapter 7 figures, Bechtel; Chapter 8 figures, Arup Environmental; Figures 9.1, 9.2, 9.3, Jean Rogers; Figures 9.4, 9.5, Jean Rogers, courtesy of Montgomery Watson; Figures 9.6, 9.7, 9.8, Jean Rogers, courtesy of Lance Larson; Figures 9.9, 9.10, 9.11, 9.12, 9.13, 9.14, 9.15, 9.16, Jean Rogers; Figure 10.1, Deborah Marton, image courtesy of Robbin Bergors; Figure 10.2, Hargreaves Associates; Figures 10.3, 10.4, Latz + Partner; Figure 10.5, Hargreaves Associates; Figure 10.6, Brown & Rowe; Figure 10.7, Arup Environmental; Figures 10.8, 10.9, Martha Schwartz Inc.; Figures 10.10, 10.11, Mike Horne; Chapter 11 Figures 11.1, 11.2, 11.5, 11.6, 11.8, 11.10, Michael Latz; Figures 11.4, 11.7, Latz + Partner; Figures 11.3, 11.9, Christa Panick; Figures 12.1, 12.2, 12.3, The Bioengineering Group; Figures 12.4, 12.5, 12.6, 12.7, Aviva Rahmani; Chapter 13 figures, William Young; Figures 14.1, 14.2, 14.3, Hargreaves Associates; Figure 14.4, Hargreaves Associates, courtesy of Presidio Archives; Figures 14.5, 14.6, 14.7, 14.8, 14.9, Hargreaves Associates; Figure 14.10, Hargreaves Associates, courtesy of Towill Inc.; Figures 14.11, 14.12, Hargreaves Associates; Figure 14.13, Hargreaves Associates, rendering courtesy of Chris Grubbs; Figure 14.14, Hargreaves Associates; Figure 14.15, Hargreaves Associates, rendering courtesy of Chris Grubbs; Figures 14.16, 14.17, 14.18, Hargreaves Associates; Figure 14.19, Hargeaves Associates, rendering courtesy of Chris Grubbs; Figure 15.1, Mike Horne; Figure 15.2, Mike Horne, courtesy of Olympic Co-ordination Authority; Figures 15.3, 15.4, Mike Horne; Figure 15.5, Mike Horne, courtesy of Olympic Co-ordination Authority; Figure 15.6, Mike Horne; Figures 15.7, 15.8, Mike Horne, courtesy of Olympic Co-ordination Authority; Chapter 16 figures, Hargreaves Associates; Chapter 17 figures, Niall Kirkwood, images courtesy of Doug Cogger, Harvard Graduate School of Design.

Bibliography

A number of the essays are followed by notes, references and further reading lists related to the particular areas of focus. Please check at the conclusions of each of the following chapters: 2 (Sattler), 4 (Carman), 6 (McNeil and Lange), 7 (Tedaldi), 9 (Rogers), 10 (Krinke), 11 (Latz), 12 (Goldsmith), 14 (Rieder), and 15 (Horne).

Further reading

Barnett, Dianna and William Browning (1995) *A Primer on Sustainable Building*, Colo.: Rocky Mountain Institute.

Carney, T. (1998) *Contaminated Land: Problems and Solutions*, New York: E. & F.N. Spon.

Craul, Philip (1999) *Urban Soils: Applications and Practices*, New York: John Wiley & Sons Inc.

Crosbie, Michael J. (1995) *Green Architecture*, Rockport, Mass.: Rockport Publishers.

Mackenzie, Dorothy (1997) *Green Design*, London: Laurence King Publishing.

Maughan, James (1993) *Ecological Assessment of Hazardous Waste Sites*, New York: Van Nostrand Reinhold.

Pilatowicz, Grazyna (1995) *Eco-Interiors, a Guide to Environmentally Conscious Interior Design*, New York: John Wiley & Sons Inc.

Raskin, Ilya and Burt D. Ensley (2000) *Phytoremediation of Toxic Metals. Using Plants to Clean Up the Environment*, New York: John Wiley & Sons Inc.

Saunders, William S. (ed.) (1998) School of Design. *Richard Haag, Bloedel Reserve and Gas Works Park*, New York: Princeton Architectural Press.

Simons, Robert (1998) *Turning Brownfields into Greenfields*, Washington, DC: Urban Land Institute.

US Green Building Council (1996) *Sustainable Building Technical Manual*, Denver: Public Technology, Inc.

Westman, Walter (1985) *Ecology, Impact Assessment and Environmental Planning*, New York: John Wiley & Sons, Inc.

Zeiher, Laura (1996) *The Ecology of Architecture*, New York: Whitney Library of Design, BPI Communications.

Index

Page numbers in *italic* refer to illustrations

Acorus calamus (sweet flag) 169
Act 2 (Land Recycling 1995) 61, 63; *see also* Pennsylvania
Ailanthus altissima (ailanthus) 156
Alisma plantago-aquatica (water plantain) 170
Alphand, J.C. 132
American Smelting and Refining Company 73
Ando, T. 115
Andropogon Associates 179
Aragon, L. 132
arsenic 72; levels 145; mud 154; production of 73, 74
Arthur Kill 133
Arup Environmental 142, 143
ASARCO Incorporated 72, 77
asbestos: abatement liability 24; hazards 16; materials containing 17

Back Bay Fens 130–1, 148
Bargmann, J. xi
Bassuk, N. 211
Bauhaus 114
Bennetton 114–15
Bestmann Green Systems, Inc. 167
bioengineering 123, 176; swale 182; techniques 175
Bioengineering Group 123, 166, 167, 171, 172, 176–7; *see also* Goldsmith, W.
bioremediation soil 43, 50
Birnbaum, C. 206
Boston: Charles River 130–1; park systems 130; British Waterways 91
Brown and Rowe 141
Brown, Capability 50
brownfields: definition 4, 61; EPA agenda 20; pilot areas

25; Louisville's Program 20; Revitalization and Environmental Restoration Act 7 *see also* Superfund
Bryant Associates, Inc. 167
Byxbee Park 139–40

Carlson, D. 1, 12–31, 252
Carman, E. xi, 33, 43–9, 50, 70, 120, 191, 246, 252
Carnegie Mellon University 63, 64; Center for Economic Development at 63; STUDIO for Creative Inquiry at 63, 64, 67; The Brownfields Center (TBC) at 63
Cephalanthus occidentalis (buttonbush) 169
Chaney, R. 240, 245
Chevron Corporation 35–40, 41, 42
Chin, M. 240
Clean Air Act 17–8, 126
Clinton, President 7, 162
coir: fiber rolls 168, *169*; woven mat 173; wrapped plugs 173
Comprehensive Environmental Response Compensation and Liability Act (CERCLA) 13, 15
Conger, K. xi, 124, 221–38, 249–50, 252
Connecticut 6, *57*
Control of Pollution Act (CPA) UK 14, 90
copper smelting operations 73
Cornus stolonifera (red osier dogwood) 169, 182
Cornell University 211
Craul, P. 141
Crissy Field 123, 133, 134–6, 148, 193–205, 206–7, 246; Golden Gate 194; Marina District 194, 199; Palace of

Fine Arts 199; Cincinatti, Ohio 38, *53*, 59
cyanide 63
Danehy Park 8, *9*
Day, J. 1, 12–31, 252
De Maria, W. 127
Denton Corker Marshall 212
Department of Energy 64
Department of Environment, UK 189, 191
detention basin 166
Detroit, Michigan *4*
diesel range organics 45
Dionne, M. 175
Disney American 18
Distichlis spicata (spike grass) 173
Dixon, R. 179
Dobson, M.C. 189
Dorchester, Massachusetts
Duisburg-Nord 123, 136–8, 150–61, 162, 248; "Cowper Place", rail harp 153, *157*; "Piazza Metallica" 150, *151*; water park 153; *see also* Latz, P.

eastern cottonwoods 56
eco-paving 247–8
Eisenman, P. 126, 148; "palimpsests" 128
Emanouil Brothers, Inc. 167
engineered phyto-treatment units 48–9
environmental audits 23
environmental expenditures: tax credits and deductions 24–6
environmental liabilities 13–18; indemnity provisions 23; insurance packages 24
Environmental Protection Act UK 14
European Community (EC) Directives 14

Exxon 133

Fabrica 115
Fagus Shoe factory 114
Federal Chemical Act (Germany) 14
Federal Medical Center, Devens 123, 166–71, 176; bioengineered vegetative installations 168–9; *see also* coir; closure 170; construction phase 169–70; participants 167; post-construction conditions 170; stormwater system 167–8
Federal Water Pollution Control Act (FWPCA) 13
Findlay, Ohio *54*
Flushing Bay 182
Forestry Authority Research Division (UK) 189
Fresh Kills landfill 5, 8, 123, 176, 178–90, 191–2; advantages 187; background 178–9; demonstration planting 184–7; garbage berms 182–4; grass and tree seeding 179–81; root penetration peport 190; salt marsh restoration 184

gabion wall 107
gas manufacturing plant *6*
gas stations 62
gasoline transfer terminal 35–7
Gasworks Park 162, 164
geomorphology 245
geostatistical analysis 72, 75–7, 246–7; nonlinear interpolation 75–6; Kriging 75–6, 79–80
Geraldton Mine Project 143–5
Glyceria canadensis (Canada manna-grass) 169
Golden Gate: General Managment Plan 196, 210; National Parks Association (GGNPA) 134, 196; National Recreation Area 196
Goldsmith, W. xi, 50, 123, 166–75, 176, 243–6, 252 *see also* Bioengineering
Grabosky, J. 211
Grand Union Canal 89, 91, 94, 95
Great Miami River 38
greenfield sites x, 8
Grimshaw, N. 115
Group, Inc 123

Haag, R. 162
Haley and Aldrich, Inc. 167
Haller, S. 195
Hanford nuclear site 164
Hargreaves, G. xi, 246, 248, 252; associates 133, 134, 135, 139, 145, 221, 222; *see also* Crissy Field; *see also* Sydney Olympics 2000
Harvard Graduate School of Design xi, xii, 123

Heizer, M. 127, 145
helixor 85, *86*
Herman Miller 114
Horne, M. xi, 123, 208–18, 246–7, 252
Hough, M. viii–ix, xi, 3, 252
Housing and Urban Development (HUD) 64

Industrial Revolution: environmental legacy 125–6
International Building Exhibition Emscher Park (IBA) 150, 151
International Erosion Control Association: environmental achievement award 167
Iris versicolor (blue flag iris) 169

Jackson, L. xi, 8, 33, 35–42, 70, 243–5, 253
Johnson C.R. Associates, Inc. 167
Juncus gerardii (black grass) 173

Kirkwood, N. 3–11, 41–2, 50–1, 59–60, 69–71, 81, 102–4, 119–21, 162, 243– 51, 253
Krinke, R. xi, 41–2, 50–1, 59–60, 69–71, 81, 102–4, 119–21, 123, 125–49, 162–5, 176–7, 191–2, 206–7, 219–20, 239–40, 253

landfill 107, 139, 146, 236; Brookfield (NY) 188, 189; cap and cover treatments 179; closures 69; site 88; gases 97–8, 143; tree roots on 178, 189; waste tipping 90; *see also* Stockley Park; *see also* Fresh Kills
landscape architects x, 245–6, 250
Lange, D. xi, 33, 61–8, 70, 102, 164, 253
Latz, A. xi,
Latz, P. xi, 71, 103, 119, 123, 136–8, 150–61, 162, 246, 248, 253
lead: abatement liability insurance 24; concentrations 72, 74; paint hazards 16
Leersia oryzoides (rice cutgrass) 169
Li, D. xi, 1, 8, 12–31, 253
Lindsay, P. 211
Lobelia cardinalis (cardinal flower) 169
Love Canal 15, 164
low temperature thermal desorption (LTTD) 197–8
Lowry Arts Collection 88

Manchester Ship Canal 83, 84, 86, 88; *see also* Salford Docks Project
manufactured sites: exhibition 125, 127, 130; landscape conference xi; locations 4; projects 133–48, post-industrial landscape 148–9;

publication ix–x; definitions 5–7
manufactured soils 247
manufactured town gas plant 6, 44
Marton, D. 243–6, 254
McDonough, W. 114
McNeil, S. xi, 33, 61–8, 70, 102, 120, 164, 246, 254
Mimulus ringens (monkey flower) 169
Minerals and Land Reclamation Division (UK) 189
Moffat A.J. 189
Montgomery Watson 197

Napa Pipe Corporation 106, 107; *see also* Oregon Steel
National Association of Waste Disposal Contractors (UK) 189
National Park Service 133, 134, 142, 193, 194, 196, 210, 250; Cultural Resource Division 195, 206; Natural Resources Division 206
National Research Council 206
National Rivers Authority 14
National Science Foundation 64
net environmental benefit analysis 40, 42
New Jersey 43
New York 4–5; Brooklyn Bridge piers; Cathedral of Saint John the Divine 171; Central Park 128; Department of Environmental Conservation (NYSDEC) 187, 188, 190; Department of Sanitation (NYCDOS) 123, 179, 184, 187, 190; greenbelt 188; harbor 182; plant climax community 184; Salt Marsh Restoration Team 133; salt marsh restoration 133–4; World's Fair (in 1939) 145; New York Times Articles 4–5, 8
Nine Mile Run 62, 64, 65, 67, 69
Northern Water Feature 124, 146–7, 221–38, 239–40; Aquatic Center pool 228, 230; construction and establishment 236; continuous deflective separation unit (CDS) 233; gross pollutant trap (GPT) 233; lemna ponds 228; Master Concept Design workshop 222–25; Master Concept Design 226, project introduction 221; "Public Domain" 222–5, 229; site description 225; technical design development 228; *see also* Sydney Olympics 2000

Ogden, Utah 35–6, 55, *56*
oil refinery 38
Olin, Laurie 128
Olmsted, F.L. 128, 130–1, 148

Oregon Steel 106, 112; *see also* Napa Pipe Corporation
Otto, F. 114
Ove Arup and Partners 83, 92; *see also* Owen R., Walker L.
Owen, R. xi, 33, 82–101, 102, 163, 249–50, 254

Panama-Pacific Exposition (1915) 195
Panicum virgatum (switchgrass) 173
Parc des Buttes-Chaumont 131–2, 138
Pennsylvania 62; Department of Environmental Protection (PaDEP) of 64; Department of Transportation (PennDOT) of 63; Southwestern Regional Planning Commission (SPRPC) of 63; *see also* Act 21 (Land Recycling 1995)
Phytokinetics Inc. 55
phytoremediation 33, 35–6, 43, *44*, 46–9, 52–8, 59–60, 176; mechanisms of 53; time in 59; hyperaccumulators in 53–4, 244–5
Phytotech Inc. 54
Pittsburgh 61, 62; technology center 63; university of 63
poplar trees 10, 36, *44*, 55
Point Defiance Park 78
Pollock and Associates 114
polyaromatic hydrocarbons (PAH) 153, 158
Populus deltoides (cottonwood) *53*
"Potentially Responsible Parties" 15, 18
Presidio, CA. 193

Rahmani, A. 166, 171, 172, 174–5, 176
regulation broker 64
remedial investigation 74
Resource Conservation and Recovery Act (RCRA) 13, 15–16
Rieder, K. xi, 123, 193–205, 243, 246, 253
RISES (Regional Industrial Site Evaluation) 64, *65*
River Irwell, Manchester 86
Roberts Harbor Tidal Marsh 123, 171–5; bioengineered vegetative installations 172; closure 174–5; construction phase 173–4; salt marsh vegetation 171; site description 171–2
Rock, S. xi, 8, 33, 52–8, 59, 81, 243–6, 254
Rogers, J. xi, 33, 105–18, 119–21, 249–51, 254
Rowe, P. xi
Ruhr Valley ix; 150, 156, *see also* Duisburg
Rutgers University 187, 191; Department of Biological Sciences 188; ecology

program 181

Saint Louis; post office 171
Salford Docks Project 82–8, 102–3, 250
Salford/Trafford Enterprise Zone 83
Salix spp. (willow) 47–8, 169, 182
Sambucus canadensis (elderberry) 169, 182
San Francisco 193; bay 196
Sattler, R. xi, 1, 8, 12–31, 255
Saw Mill Creek 133–4
Scarpa, A. and T. 115
Schwartz, M. 143, *see also* Geraldton Mine Project
Scirpus validus (soft-stem bulrush) 169
Shepard Epstein and Hunter 83
Silicon Valley 115 *see also* Superfund
Silverstein, M. 105
Smithson, R. 126–7, 148; earthworks 127–8, 140
Spartina alterniflora (smooth cordgrass) 133, 173
Spartina patens (salt meadow grass) 173
Spectacle Island 139, 140–2
Staten Island 123, 133, 178; greenbelt 184
Stockley Park 88–101, 102–3, 139, 142–3; business park 92–4; community and business 100; county park 95, 96, 98, 99; management of landfill gases 97–8; planting and seeding 97; success of reclamation 99–100
St. Paul, Minnesota 240
Sudbury, Ontario ix
Superfund 20–2, 106, 113; lender liability 21; Memorandum of Agreement 20; potential purchaser agreements 21; risk-based clean-up costs 21
Sydney Olympics 2000 123, 145–8, 208–18, 219–20; background 208–10; eco-paving 214–16, 219; Government Architect Design Directorate (GAAD) 208; Haslams Pier 208, 210, 217; Metropolitan Meatworks 210; Millennium Parklands 208; Olympic Co-ordination Authority 208, 217; Olympic Plaza 147, 148, 208, 210, 214, 219, 220; plaza figs 210; structural soil 210–14; *see also* Northern Water Feature; *see also* Horne, M.

Tacoma Asarco smelter site 70, 72–80
Tacoma Yacht Club 78–9
Tedaldi, D. xi, 33, 70, 81, 246–7, 255
Thames Water 91

Tonkin Zulaikha, 214
Toronto viii, port lands ix
total petroleum hydrocarbon (TPH) 133
Toxic Substances Control Act (TSCA) 16–17
Trenton, New Jersey 54
trichloroethene (TCE) 56

Ukeles, M. 176
underground storage tanks (UST's) 16; insurance of 24
Urban, J. 219
Urban Redevelopment Authority (URA) 64, 67
US Air Force 56
US Army, 194, 206
US Congress 13, 25
US Department of Agriculture 240
US Environmental Protection Agency (US.EPA) 14, 35, 38, 42, 64, 72, 76, 190, 249; National Risk Management Laboratory 59; Superfund List 72
US Green Building council 114
US Inland Revenue Service 25
US Soil Conservation Service (SCS) 186

Viburnum dentatum (arrowwood viburnum) 169
Vinalhaven, Maine 166, 176; oatmeal quarry 171; *see also* Roberts Harbor tidal marsh
Voelklingen 156

Walker, L. xi, 33, 82–101, 102, 163, 249–51, 255
Washington's Landing 62, 64, 65, *66*, 69
Wastes Technical Division, UK 189
waterfront renewal conferences viii
Water Resources Act, UK 14
Wells National Estuarine Research Reserve, Maine 175
West Shore Expressway 182
wetlands 126; construction 166–75; plantings 169–70, 171; *see also* Goldsmith, W. *see also* Northern Water Feature
Wexner Center for the Visual Arts 128–9
Wilkhahn 114
Winterbottom, D. xi, 41–2, 50–1, 59–60, 69–71, 81, 102–4, 119–21, 162–5, 176–7, 191–2, 206–7, 219–20, 239–40, 249–50, 255
Wisconsin 44, Milwaukee 49

Young, W. xi, 8, 123, 178–90, 191–2, 243, 245, 255

Zaitzevsky C. 130–1